Synthesis Lectures on Visualization

Series Editors

David Ebert, University of Oklahoma, Norman, USA

Niklas Elmqvist, College of Information Studies, University of Maryland, College Park, MD, USA

This series publishes on topics pertaining to scientific visualization, information visualization, and visual analytics. Potential topics include, but are not limited to scientific, information, and medical visualization; visual analytics, applications of visualization and analysis; mathematical foundations of visualization and analytics; interaction, cognition, and perception related to visualization and analytics; data integration, analysis, and visualization; new applications of visualization and analysis; knowledge discovery management and representation; systems, and evaluation; distributed and collaborative visualization and analysis.

Shixia Liu · Weikai Yang · Junpeng Wang ·
Jun Yuan

Visualization for Artificial Intelligence

 Springer

Shixia Liu
School of Software
Tsinghua University
Beijing, China

Weikai Yang
The Hong Kong University of Science
and Technology (Guangzhou)
Guangzhou, Guangdong, China

Junpeng Wang
Visa Research
Foster City, CA, USA

Jun Yuan
School of Software
Tsinghua University
Beijing, China

ISSN 2159-516X   ISSN 2159-5178   (electronic)
Synthesis Lectures on Visualization
ISBN 978-3-031-75339-8   ISBN 978-3-031-75340-4   (eBook)
https://doi.org/10.1007/978-3-031-75340-4

This Springer imprint is published by the registered company Springer Nature Switzerland AG
The registered company address is: Gewerbestrasse 11, 6330 Cham, Switzerland

If disposing of this product, please recycle the paper.

Acknowledgements

We are deeply grateful for the contributions and support from a wide range of individuals.

First of all, we would like to express our sincere thanks to Prof. Xiting Wang, Prof. Changjian Chen, and Prof. Jing Wu for their insightful discussions and invaluable feedback throughout the development of this book. Their expertise and constructive suggestions have been pivotal in shaping the final manuscript. We also appreciate the thoughtful suggestions provided by the anonymous reviewers, which greatly improved both the content and presentation of this book. Special thanks go to Duan Li for his insightful suggestions and efforts in improving the figures.

We are grateful to Prof. Niklas Elmqvist and Prof. David Ebert for offering us the opportunity to publish this book in their synthesis lecture series on visualization. Niklas' continuous encouragement, feedback, and support have been crucial in bringing this book to fruition. We also extend our appreciation to Christine Kiilerich, Boopalan Renu, and Jayanthi Narayanaswamy for their excellent work in managing and editing this book.

Our research was supported by the National Natural Science Foundation of China under grants U21A20469 and 61936002, whose financial support was essential for this work.

Finally, we would like to acknowledge the unwavering encouragement, patience, and support of our friends and families. Their belief in this book has been a constant source of strength.

November 2024
Shixia Liu
Weikai Yang
Junpeng Wang
Jun Yuan

Contents

Introduction

Artificial intelligence (AI) refers to the capability of a computer to perform cognitive tasks that are typically associated with human intelligence such as learning, reasoning, and problem-solving. Due to the success of machine learning, especially foundation models (e.g., ChatGPT), the field of AI is currently experiencing rapid growth and has the potential to revolutionize many aspects of our lives, from health care to finance, criminal justice to education. With the increasing deployment of AI systems in these fields, ensuring their ability to generalize to new, unseen data has become crucial for optimal performance and practical utility in real-world applications. Furthermore, it is crucial that the decisions made by these systems are transparent and explainable for greater accountability and trustworthiness.

1.1 Generalization and Interpretability of AI

Machine Learning serves as the foundational backbone and a critical component in the field of AI. It centers on the development and refinement of models that learn from large amounts of data, identify patterns, predict outcomes, and make decisions based on input data. Generalization, the ability of an ML model to perform well on new and previously unseen data, is crucial for high performance and real-world applicability. This ability enables an AI system to perform robustly in the presence of changes in the data distribution. When a model lacks strong generalization capabilities, it tends to overfit the training data by capturing the spurious correlations and statistical noise of the training data instead of the underlying patterns and rules. In such cases, the model may perform poorly on new data and result in poor performance and low applicability in real-world scenarios. Therefore, ensuring that AI systems can generalize well is a critical challenge in the development and deployment of these systems.

© The Author(s), under exclusive license to Springer Nature Switzerland AG 2025   1
S. Liu et al., *Visualization for Artificial Intelligence*, Synthesis Lectures on Visualization,
https://doi.org/10.1007/978-3-031-75340-4_1

On the other hand, as AI becomes more ubiquitous in various high-stake tasks such as precision medicine, law enforcement, and financial investment, there is a growing need for transparency and interpretability of ML models and their predictions. This is where explainable artificial intelligence (XAI) comes in [8, 146]. It enables users to understand the inner workings of these models and trust their generated outputs. In the aforementioned areas, the significance of XAI has increased substantially due to the potential risks and consequences related to inaccurate or biased predictions. This makes XAI techniques essential to ensure the reliability, fairness, and accountability of AI systems [132]. First, the increasing reliance on machine learning models, particularly complex ones like large language models, has made the field of XAI indispensable. These models may have billions of parameters or even more, which makes them hard to interpret. This lack of transparency raises trust issues, especially in critical applications. For example, doctors may be reluctant to rely on a deep model for diagnosing medical conditions if it cannot explain its predictions. Second, without understanding how a model works, it can be difficult to diagnose problems when the model fails to perform as expected. For example, if a self-driving car fails to detect an obstacle and causes an accident, it is important to understand why the associated model failed. This understanding is crucial for implementing effective corrections. Third, XAI is essential to ensure fairness and accountability in decision-making. Machine learning models are increasingly used to make decisions that have a significant impact on people's lives, such as determining whether to grant a loan or hire a job candidate. If the model is biased or unfair, it can have a negative impact on certain groups of people. XAI can help ensure that models are fair and unbiased by providing insights into how the models arrive at their decisions. Finally, XAI can help build trust and acceptance of machine learning models in real-world applications. People are often skeptical of models they do not understand, and this can lead to resistance or even rejection of the technology. By providing explanations of how models work, XAI fosters trust and acceptance among users, which leads to more widespread adoption of machine learning technologies.

1.2   Visualization for AI

Visualization transforms data into graphical forms such as charts, graphs, and maps, and allows users to interact with it. Its interactive nature enables users to engage directly with the data, often in real time. This interaction results in a more dynamic and personalized understanding of the data presented. Visualization is frequently used in data analysis and decision-making activities because it allows users to examine complex and massive data from various angles and gain insights and understanding that may not be obvious through other methods. It has demonstrated effectiveness in providing explanations, facilitating communication, and promoting human-machine collaboration [153]. This makes visualization a suitable choice for fully understanding and analyzing AI systems [259, 270]. Visualization can be particularly useful for understanding the data used to train machine learning models, the inner workings of these models, and how they arrive at their predictions. Consequently, the area of visualization for artificial intelligence (VIS4AI) has emerged as an exciting area for research and development. It offers many opportunities for advancing the use of visualization techniques that can improve the interpretability and reliability of machine learning models. In addition, VIS4AI fosters human-machine collaboration and paves the way for more reliable and effective AI applications.

VIS4AI methods fully combine the advantages of interactive visualization and machine learning techniques to facilitate the analysis and understanding of key components in the learning process, with the aim of improving performance. For example, research in VIS4AI that focuses on explaining the inner workings of deep neural networks has successfully increased the transparency of deep models and has received growing attention from the research community in recent years [44, 98, 154, 272]. These methods are applicable in all phases of the machine learning lifecycle, from data preparation to model development.

As shown in Fig. 1.1, the machine learning lifecycle consists of two pipelines: a data pipeline and a model pipeline. The **data pipeline** prepares the data for machine learning. It typically includes data collection, data cleaning, data augmentation, and feature engineering. The goal of this pipeline is to ensure that the training data collected is representative, unbiased, and of high quality, while also ensuring that the training data contains the necessary features for training a machine learning model. A well-designed data pipeline guarantees not only the accuracy and robustness of the model, but also its ability to effectively generalize to new datasets. The **model pipeline** involves selecting, training, evaluating, and deploying a machine learning model. This pipeline consists of both the model development and model deployment stages. Model development centers on the creation, training, and optimization of a machine learning model. This includes tasks such as model selection, training, validation, and evaluation, all of which aim to identify the best model for the given task and improve its performance. Model deployment aims to make a trained machine learning model available for use in a production environment. This stage includes the tasks of monitoring the model such as scalability, reliability, security, and fairness, as well as maintaining the model such as addressing concept drift. A well-designed model pipeline ensures that the model is accurate, reliable, and robust and, therefore, capable of delivering effective and trustworthy results.

Both the data pipeline and the model pipeline are important components of machine learning. They are often used in combination to build and deploy effective machine learning models. By carefully designing and implementing each pipeline, machine learning practi-

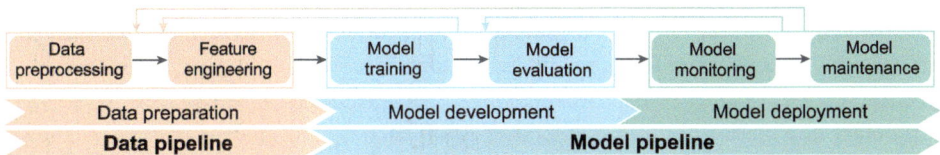

Fig. 1.1 Machine learning lifecycle consists of a data pipeline and a model pipeline

tioners can ensure that their models are accurate, robust, and effective in solving real-world problems. VIS4AI methods seamlessly integrate with the well-established data pipeline and the model pipeline essential for developing and deploying machine learning models. In the data pipeline, VIS4AI techniques aim to improve the quality of data and features used to train machine learning models. This includes tasks such as cleaning training data and creating interpretable and meaningful features through feature engineering. VIS4AI techniques can effectively identify and mitigate issues such as dataset bias, annotation inconsistency, and outliers, which affect the accuracy and reliability of machine learning models. In the model pipeline, VIS4AI techniques support the development and deployment of machine learning models. During model development, these techniques facilitate model understanding, diagnosis, and steering. They employ visualizations to explore and understand the behavior of the model, identify potential issues, and suggest improvements. Once the model has been developed, the VIS4AI techniques assist in model deployment by enabling decision explanation, model monitoring, and model maintenance. They use interactive visualization techniques to explain the model decisions, monitor model performance in real time, and maintain the performance by tackling the robustness and fairness issues.

1.3   The Development of VIS4AI

Over the past two decades, there has been a growing interest in the development of VIS4AI techniques. The goal of these techniques is to enhance the generalizability, interpretability, trustworthiness, and reliability of machine learning models by using visualization techniques. This has become increasingly important as machine learning continues to be used in various applications. To achieve this goal, many VIS4AI methods have been developed to advance the understanding of how an ML model works and how it arrives at its predictions. Figure 1.2 summarizes the evolution of VIS4AI methods over time. The lower part outlines the VIS4AI methods related to the data pipeline, and the upper part outlines the VIS4AI methods related to the model pipeline. The VIS4AI methods on the model pipeline can be classified into two categories based on their target users: model development tailored for model developers and model deployment designed for model consumers.

Initial attempts in VIS4AI for the data pipeline focus on feature engineering, which involves interactive feature selection [124] and feature creation [24]. Later, there are increasing advocates for improving the quality of training data, including improving crowd-sourced annotations [157], correction of mislabeled data [260], and the detection of out-of-distribution samples [38]. The researchers also proposed to use unannotated data [37] and multi-modal data [39] to improve model performance. Recently, efforts have been made to reweight training samples in order to address data bias, including noisy labels and imbalanced class distributions in training datasets [267]. In addition, how to detect and analyze data heterogeneity in federated learning is also studied [251].

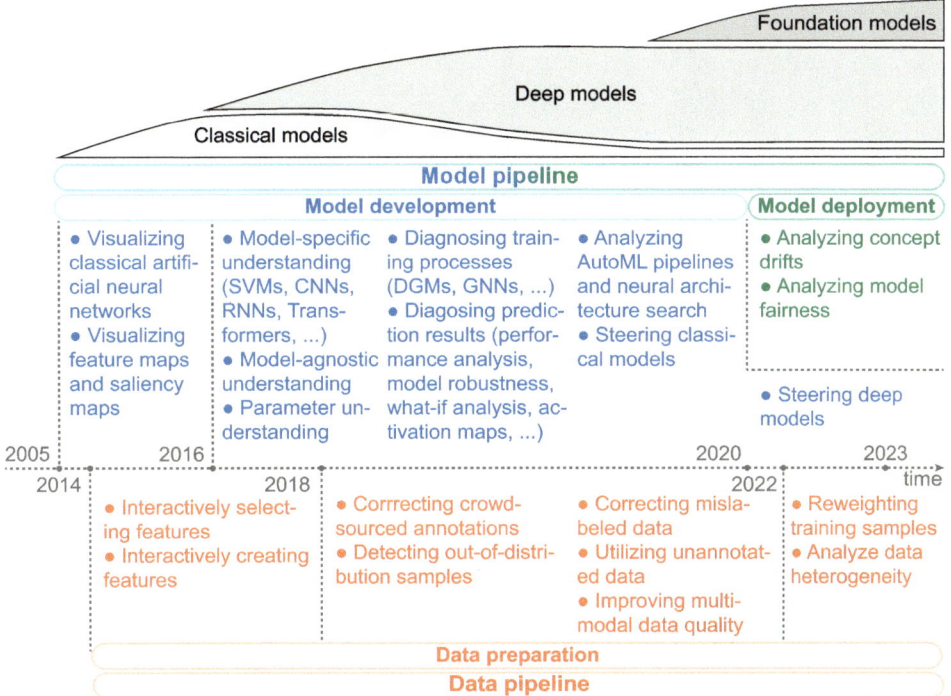

Fig. 1.2 The evolution of VIS4AI methods over time

Initial efforts in model development focus on utilizing visualization techniques to facilitate the understanding of classical models such as neural networks [234], decision trees [232], and regression models [178]. With the development of deep models, efforts have been shifted to understanding various deep models such as convolutional neural networks (CNNs) [154], recurrent neural networks (RNNs) [172], deep generative models (DGMs) [115], and transformer-based models [53, 145]. Hereafter, researchers seek to diagnose the training process of machine learning models [151] and the prediction results [27]. Recent efforts focus on analyzing AutoML pipelines [186] and neural architecture search [271], and steering classical models [264] and deep models [175]. In the model deployment stage, the researchers focused on analyzing concept drifts [263] and model fairness [26].

1.4   Conceptual Framework and Method Overview

The core principle of VIS4AI is based on the well-established mantra of visual analytics, which advocates the integration of interactive visualization and data analysis techniques to facilitate human reasoning and decision-making processes. Keim et al. [119] defined visual

Fig. 1.3 A conceptual framework of VIS4AI adapted from the visual analytics process proposed by Keim et al. [119]

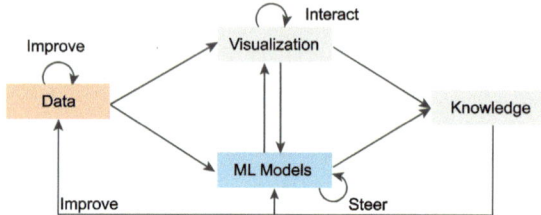

analytics as the integration of automatic analysis techniques with interactive visualizations to facilitate effective understanding, reasoning, and decision-making based on large and complex datasets. The inherent ability of visual analytics to provide explanations, facilitate communication, and offer intuitive insights makes it a powerful tool to support VIS4AI. The conceptual framework adapted from the visual analytics process proposed by Keim et al. [119] is shown in Fig. 1.3. It illustrates the use of the visual analytics process to understand the impact of each phase on the final predictions. Specifically, we focus on introducing the VIS4AI-related process, which includes improving the quality of training data, facilitating the development of ML models, and supervising the deployment of ML models into production environments (Fig. 1.4).

The quality of training data is the main factor in the success of machine learning, especially with respect to the model capability to generalize and maintain robustness [47, 157]. It is essential to note that regardless of the complexity of the learning model or the amount of data used to train it, the upper limit of its performance is constrained by the quality of the training data [47]. However, obtaining high-quality training data can be an extremely

Fig. 1.4 VIS4AI techniques are widely used in both the data pipeline and the model pipeline. These techniques enable effective visual analysis of complex data and the interpretation of machine learning models

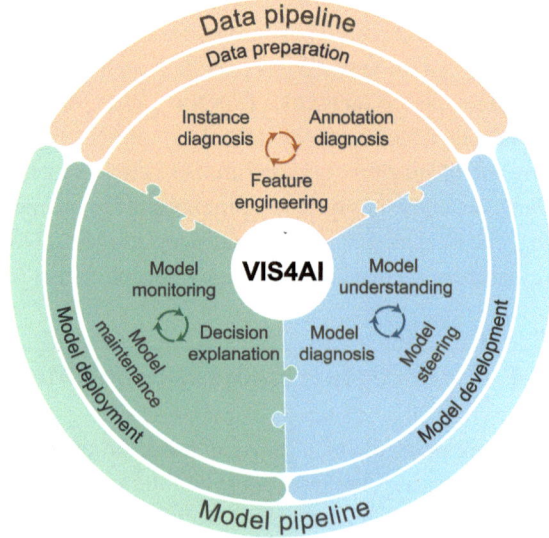

time-consuming and expensive process. Recent studies have shown that improving data quality consumes approximately 30–80% of time and resources in data-driven analysis and decisions [121, 182, 215]. Such a huge demand for human effort leads to increased research focused on improving training data in the field of visual analytics [266, 270]. VIS4AI methods, which tightly integrate interactive visualization with machine learning techniques, have emerged as a promising solution to reduce human effort and improve the quality of training data. The corresponding efforts are classified into three categories: instance diagnosis, annotation diagnosis, and feature engineering. Instance diagnosis focuses on identifying and addressing specific issues at the data instance level, including missing values, out-of-distribution (OoD) instances, and ambiguous instances with difficult-to-recognize content. Annotation diagnosis is primarily concerned with annotation-level issues and addresses the issues of inaccurate annotations, insufficient annotations, and inexact annotations. Feature engineering aims to improve the performance of ML models through the addition of crucial features and the elimination of unnecessary ones.

Model development is the secondary factor in building a successful AI application. It involves selecting the most suitable model architecture for the task at hand and training it on the available data. However, model development is often a challenging and time-consuming task, especially when working with large and complex datasets. For example, when the training process fails or the model does not provide satisfactory performance, model developers need to diagnose the issues that occur in the training process. Therefore, developing VIS4AI methods to explain the inner workings of machine learning models and diagnose the issues that occurred in the model building process has become a growing research direction in visualization. These methods can better illustrate the inner workings of the model training process and significantly improve the efficiency and effectiveness of building machine learning models. In this book, we categorize current methods by their analysis tasks: model understanding, diagnosis, and steering. Model understanding aims to visually explain the inner workings of a model, such as how changes in parameters influence the model and why the model gives a certain output for a specific input. Model diagnosis focuses on the diagnosis of errors in model training through interactive exploration of the training process. Model steering is primarily focused on interactively improving model performance. For example, to refine a topic model, Utopian [45] enables users to interactively merge or split topics and automatically modify other topics accordingly.

The deployment of a trained model into a production environment is another critical phase toward a successful AI application. This phase aims to ensure transparency, trustworthiness, fairness, and robustness, which are essential to maintain the integrity and practicality of AI applications in real-world scenarios. In this phase, the VIS4AI methods play an important role in bridging the gap between the intricate internal mechanisms of the model and the practical demands of its users. These methods are categorized into two main classes: decision explanation and model monitoring and maintenance. Decision explanation explains the intricate process of interpreting the pathways through which a model arrives at its decisions.

This aspect of model deployment not only illuminates the inner workings of the model but also significantly enhances transparency. By clarifying the decision-making process, it fosters a deeper understanding and consequently strengthens users' confidence and trust in the model capabilities. On the other hand, model monitoring and maintenance focus on the ongoing attention required to maintain model effectiveness over time. This involves not just ensuring consistent performance, but also protecting against biases, ensuring fairness, and adapting to evolving data landscapes to maintain robustness. These tasks are particularly challenging given the dynamic nature of real-world applications, where data and the environment are constantly in flux.

1.5 Book Motivation and Structure

1.5.1 Book Motivation

This book provides a comprehensive framework for VIS4AI and introduces the associated techniques in the data and model pipelines. It emphasizes the importance of interactive visualization in AI and presents a variety of VIS4AI techniques for different purposes. In addition, it discusses the challenges and opportunities of VIS4AI and proposes several promising research topics for future work such as improving training data using complementary modalities, online training diagnosis, and unified model evaluation with interactive visualization. Overall, this book is a resource for researchers and practitioners who are interested in visualization and artificial intelligence.

1.5.2 Book Structure

Chapter 2 sets the foundation for VIS4AI by covering essential elements such as data types and machine learning models. It introduces commonly used data types such as tabular data, sequential data, multi-dimensional array, graph, and multi-modal data, along with widely adopted machine learning models such as support vector machines (SVMs), decision trees, convolutional neural networks (CNNs), recurrent neural networks (RNNs), and generative pre-trained transformers (GPTs).

Chapter 3 focuses on VIS4AI methods and techniques for data preparation, including diagnosing instances, diagnosing annotations, and performing feature engineering.

Chapter 4 covers the VIS4AI methods and techniques for model development, including model understanding, model diagnosis, and model steering.

Chapter 5 presents the VIS4AI methods and techniques for model deployment, including decision explanation, model monitoring, and model maintenance.

Chapter 6 explores various research challenges and opportunities faced by VIS4AI, including evaluation, human factors, foundation models, data-centric AI, and model-agnostic explanation.

Finally, the backmatter provides a comprehensive summary and conclusion.

Fundamentals

Since VIS4AI techniques are widely used in both the data and model pipelines, this chapter provides a comprehensive overview of prevalent data types and machine learning models integral to these pipelines. We cover a range of data types, including tabular, sequential, multi-dimensional array, graph, and multi-modal data. On the model front, we explore various popular models categorized into classical models, deep learning models, and foundation models. In the category of classical models, we introduce linear/logistic regressions, decision trees and tree ensembles, and support vector machines (SVMs). Our investigation into deep learning models includes multi-layer perceptrons (MLPs), convolutional neural networks (CNNs), recurrent neural networks (RNNs), graph neural networks (GNNs), deep generative models (DGMs), and transformers. Additionally, we explore advanced foundation models such as bidirectional encoder representations from transformers (BERT), vision transformer (ViT), InternImage, contrastive language-image pertaining (CLIP), and the generative pre-trained transformer (GPT) series models.

2.1   Data

Based on a comprehensive review of the existing VIS4AI literature, we categorize training data into the following five types: tabular data, sequential data, multi-dimensional array data, graph data, and multi-modal data. Each sample dataset consists of a set of instances, often accompanied by relevant annotations. The sample dataset $\mathcal{D}$ can be described as

$$\mathcal{D} = <X, Y>, \ where$$
$$X = \{x_1, x_2, ..., x_n\} \ and \ Y = \{y_1, y_2, ..., y_n\}. \tag{2.1}$$

X is the feature vectors of all the instances in $\mathcal{D}$, which is the input of ML models. A feature in the feature vector is a measurable attribute or characteristic of an observed phenomenon [34]. Examples include (1) the age, gender, or annual income of an individual; or (2) the color,

11
S. Liu et al., *Visualization for Artificial Intelligence*, Synthesis Lectures on Visualization,
https://doi.org/10.1007/978-3-031-75340-4_2

edge, or texture of an image. If Y exists, it is the annotation set that the supervised or semi-supervised learning models aim to learn during training. This could be, for example, the class labels of image data or the bounding boxes that delineate different objects in images. Each sample in $\mathcal{D}$ is denoted as (x_i, y_i). If the dataset $\mathcal{D}$ lacks the Y set, machine learning models resort to unsupervised or self-supervised learning methods based on the X set.

The differences between the five data types reside in the X set. We explain them by (1) providing their definition, (2) listing some typical examples, and (3) discussing the challenges when learning from them.

2.1.1 Tabular Data

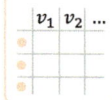

Definition. The tabular data comes as a data table, where each row is a sample, and each column is a feature of the sample. Accordingly, a sample is denoted as

$$x_i = (v^{D_1}, v^{D_2}, ..., v^{D_m}),\tag{2.2}$$

where v^{D_j} ($j \in [1, m]$) is a possible value of the jth feature defined in the corresponding domain D_j, which can be categorical or numerical. If the Y set exists, it usually appears as a column in the table.

Examples. The U.S. Census Income dataset used in the What-If Tool developed by Wexler et al. [254] is a typical tabular dataset. Each row of the dataset is a person, and each column represents the value of one feature, such as age, gender, and capital gain. Similar examples also include the Bank Marketing dataset used in RuleMatrix [173], and the Criminal Recidivism dataset used in CoFact [118]. The individual features of these tabular data are usually human-understandable, which contributes significantly to the interpretation of the corresponding ML models. Moreover, new features can also be generated through feature engineering such as the combination of two features.

2.1.2 Sequential Data

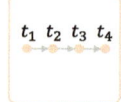

Definition. Sequential data comes with a collection of sequences that have a specific order and are often of varied lengths. The order of the sequences is important and usually represents a temporal or spatial progression. Each sequence x_i consists of k different elements, and each element is represented by a feature vector. For example, a sentence with k words is a sequence of k elements. Each element e_i is an embedding vector of a word. Typically, x_i is defined as

$$x_i = (e_1, e_2, ..., e_k). \tag{2.3}$$

Examples. The two most common sequential data are textual data (each word/character is a token) and time-series data (each time step is a token). For example, the Penn TreeBank [167] dataset used in RNNVis [172] and LSTMVis [227] is a famous English corpus of sentences. Each sentence is a sequential sample, and the parts of speech for individual words/tokens have been well-annotated in the dataset. Weather forecasting data [208], sleep signals [77], and musical chord progression sequences [227] are examples of time-series data, in which tokens are ordered chronologically into sequences.

2.1.3   Multi-dimensional Array Data

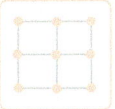

Definition. Multi-dimensional array data is composed of a set of samples. Each sample is an array of scalar values organized spatially into a regular grid structure. For example, a gray-scale image can be considered as a 2D array storing the image's pixels along the width and height dimensions. In a 2D space, each multi-dimensional array sample is denoted as

$$x_i = \begin{pmatrix} s_{1,1} & s_{2,1} & \cdots & s_{w,1} \\ s_{1,2} & s_{2,2} & \cdots & s_{w,2} \\ \vdots & \vdots & \ddots & \vdots \\ s_{1,h} & s_{2,h} & \cdots & s_{w,h} \end{pmatrix}.$$

Examples. Image and volume data are representative examples of this type of data. For example, the MNIST dataset [52] used in GANViz [239] and Rauber et al.'s work [203] is a famous benchmark that consists of 70,000 gray-scale images of hand-written digits. Each image is a 2D array with individual scalar values (pixels) ranging from 0 to 255. The CIFAR10 [126] dataset used in CNNVis [149] and DGMTracker [151], and the ImageNet dataset [212] used in Summit [99] and Bluff [50] are high-resolution RGB image datasets with many classes. In these datasets, each image is a 3D array of scalars. Additionally, 3D volume data generated from scientific simulations or medical imaging have also been extensively used for the training of machine learning models. For example, the combustion and ionization datasets used in SSR-TVD [88] comprise 3D volumes, with each grid point storing a scalar value corresponding to the respective physical property. Similarly, the CT-Chest dataset used in DNN-VolVis [100] serves as another illustration of volume data derived from the field of medical imaging.

2.1.4 Graph Data

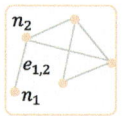

Definition. A graph consists of a set of nodes and edges. Nodes are individual entities in the graph, and edges represent relationships between entities. Formally, a graph is defined as follows:

$$\mathcal{G} = \langle \mathcal{N}, \mathcal{E} \rangle,$$
$$\text{where } \mathcal{N} = \{n_1, n_2, ..., n_n\}, \mathcal{E} = \{e_{i,j} \mid 1 \leq i, j \leq n\}. \tag{2.4}$$

Each node is further represented by a feature vector:

$$n_i = (f_{i1}, f_{i2}, ..., f_{ik}). \tag{2.5}$$

Graphs are typically divided into two categories: homogeneous graphs and heterogeneous graphs. In homogeneous graphs, all graph nodes are of the same type, and all graph edges denote the same relationship between nodes. In contrast, heterogeneous graphs consist of nodes of different types, and the edges can represent multiple types of relationships.

Examples. A social network is a typical homogeneous graph, where each node is a person, and each edge encodes the friendship between persons. Each person is represented by multiple features such as gender, age, and the number of friends. These constitute the feature vector of the corresponding graph node. Other homogeneous graph examples include publication citation graphs [168] and molecular compound structure graphs [177]. In the case of heterogeneous graphs, the User-Movie data used in CorGIE [159] is a typical example, where a graph node can be a user or a movie, and an edge between two nodes represents that the user has watched the corresponding movie. It is important to note that there are also ML models designed to learn from a set of graphs. Here, each training sample is a graph rather than a graph node. These individual graphs have independent sets of nodes and edges. For example, a chemical compound can be represented as a graph where nodes are atoms and edges are bonds. Researchers have developed many deep models to predict whether a compound is cancer-related or not [238].

2.1.5 Multi-modal Data

Definition. Multi-modal data includes data from multiple sources and types that capture different aspects of a subject. A modality represents a unique form of data representation, such as text, audio, video, and sensor readings. The arrangement of these modalities can be nested or collocated. In the nested arrangement, one modality predominates the data format, with other modalities embedded within it. In contrast, the collocated arrangement features multiple modalities of equal significance with a lower degree of interconnection. Typically, the integration of these diverse modalities can enhance the performance of machine learning models and provide a more comprehensive understanding of the subject under study.

Examples. Video data can be considered a hybrid of multi-dimensional array data and sequential data. Each frame of the video is an image that encodes spatial features. A consecutive sequence of these frames constitutes sequential data. The spatial modality is nested within the sequential modality. Most deep reinforcement learning (DRL) agents trained to play video games use this type of multi-modal data as training data such as the game's episodes used in DRLive [244] and DRLViz [107]. Dynamic graphs combine sequential data with graph data, and the graph modality is nested under the sequential modality. An evolving social network is such an example in which the number of nodes (users) and edges (users' relationships) keep changing over time. Different modalities can also be collocated at the same level. For example, the data used in M^2Lens [250] include three types of sequential data with different modalities: (1) facial expressions (video data), (2) voices of speakers (acoustic data), and (3) verbal transcripts (textual data). Different AI models can be trained to take care of the respective modalities, and their outputs can be fused together for comprehensive learning.

2.2 Machine Learning Models

We categorize machine learning models into three groups: classical models, deep models, and foundation models. The model size increases steadily across these groups. Accordingly, as indicated by the vertical axis in Fig. 2.1, their performance tends to improve from the earlier to the later groups. However, enhanced performance often sacrifices explainability, as highlighted by the horizontal axis of the same figure. This section provides a brief overview of the major machine learning models within each group, which supports subsequent discussions on their explainability via VIS4AI. For detailed information on individual models, the reader can refer to references [20, 80].

2.2.1 Classical Models

Classical models, often based on mathematical and statistical principles, have relatively fewer parameters. Typical examples include linear regressions, logistic regressions, decision trees, tree ensembles (e.g., random forests, boosting trees), and SVMs.

Linear/Logistic Regressions. Linear regressions are statistical models that fit the linear relationship between a dependent variable y and a set of independent variables $X = \{x_1, x_2, \ldots, x_n\}$: $y = a_0 + \sum_{i=1}^{n} a_i x_i$. Here, a_0 is the y-intercept. The coefficient a_i of the variable

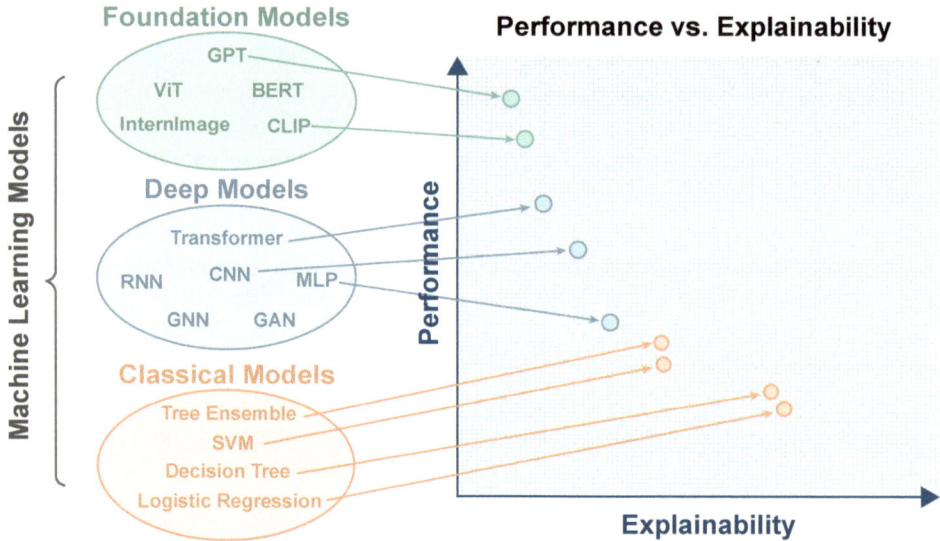

Fig. 2.1 Learning performance versus explainability trade-off for different machine learning models. This is inspired by DARPA's work [86]

x_i ($i \in \{1, 2, \ldots, n\}$) reflects its contribution to reaching y. The model can be trained by tuning the values of a_i ($i \in \{0, 1, \ldots, n\}$), so that the result of $a_0 + \sum_{i=1}^{n} a_i x_i$ is as close to y as possible. Logistic regressions work similarly to linear regressions but are tailored for classification tasks. Unlike linear regressions, they assume that the target variable y is a binary categorical variable. The probability of y belonging to the positive class is modeled by applying the sigmoid function $g(v) = 1/(1 + \exp(-v))$ to the linear combination $a_0 + \sum_{i=1}^{n} a_i x_i$.

Decision Tree and Tree Ensembles. The decision tree is a supervised machine learning model used for both classifications and regressions. Take classification as an example; the model is usually built by first taking a feature and a condition of the feature that can split data samples into subgroups, such that the samples in each group are as pure as possible with respect to labels. The purity level of the samples can be measured by various metrics such as the Gini index and information gain. This procedure is repeatedly applied to the sample subgroups to build the decision tree until certain termination criteria are met, such as reaching the tree's maximum depth or the minimum number of samples in each leaf node. To use this decision tree for inference or prediction, one simply traverses the tree from top to bottom. The prediction result is the majority class of samples found in the leaf node reached. Figure 2.2 shows an example of a decision tree used to predict indoor or outdoor activities. The model prediction is to play outside if the weather is sunny and the wind is weak, as indicated by the red arrow.

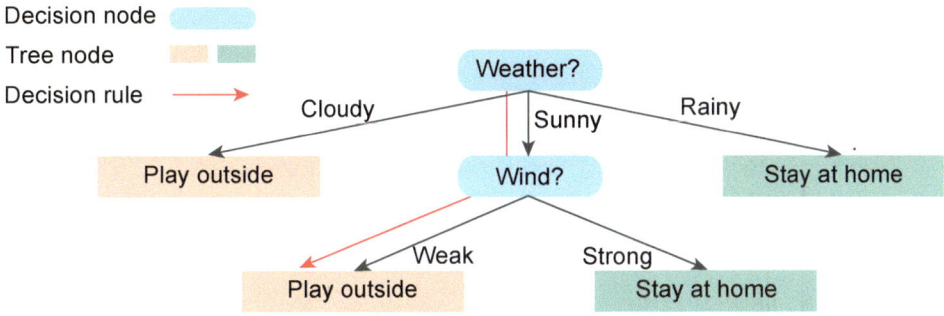

Fig. 2.2 An example of a decision tree for predicting indoor or outdoor activities

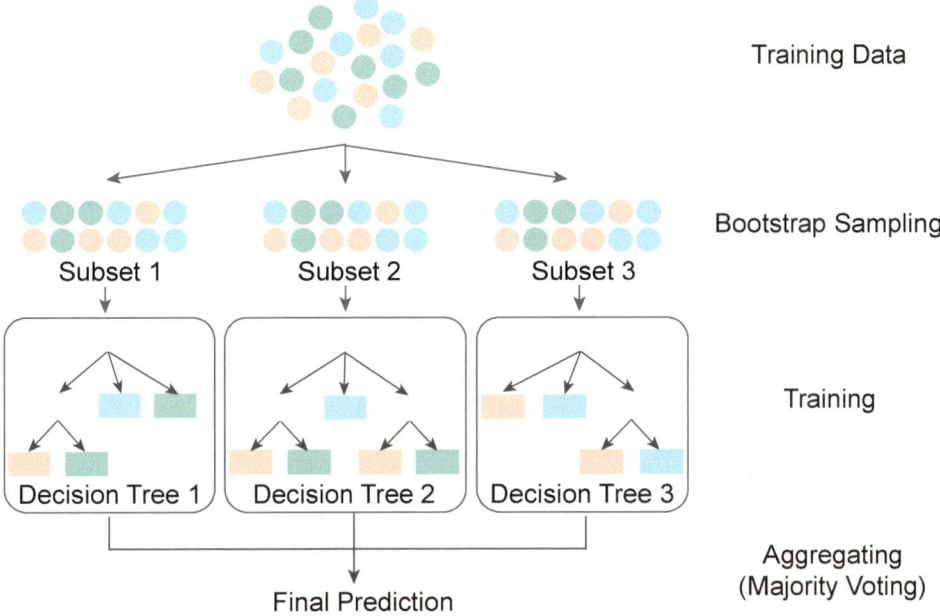

Fig. 2.3 An illustrative example of building a bagging model

Trees can be ensembled in two main ways, including bagging and boosting, to improve their prediction capabilities. As shown in Fig. 2.3, bagging independently trains multiple models on different subsets of the training data. These subsets are created through bootstrap sampling, where data samples are randomly selected with replacement. This sampling with replacement strategy selects a data sample from the training dataset, and then puts it back to the dataset before the next selection, allowing it to be potentially selected again. This ensures that the base models are trained on diverse subsets of the training data, as some samples may appear multiple times in the new subset, while others may be omitted, effectively reducing

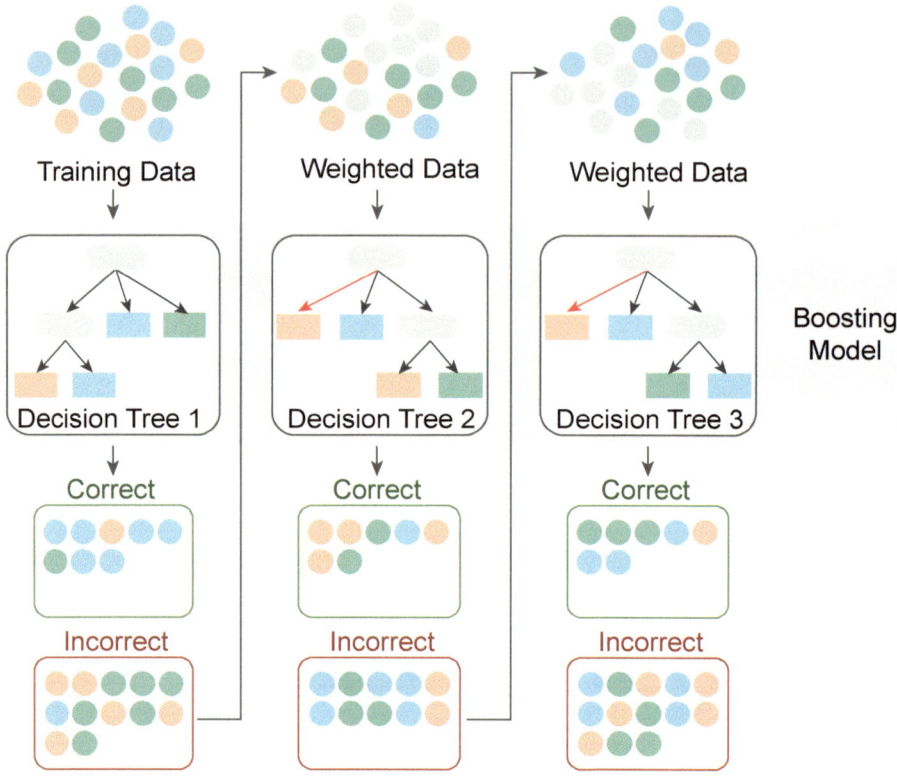

Fig. 2.4 An illustrative example of building a boosting model

the risks of overfitting and improving the accuracy of the model. The final prediction is made by averaging the predictions from all models for regression, or by majority voting for classification. The random forest is a typical example of the bagging ensemble method. Boosting is an ensemble learning technique that builds a series of models in a sequential manner, where each subsequent model aims to correct the errors made by the previous ones. Figure 2.4 illustrates the process of building a tree ensemble based on the boosting strategy. Boosting trees use all training samples to build an initial decision tree. Each subsequent decision tree is built with a focus on the samples that are incorrectly classified by the previous trees, by assigning higher weights to these misclassified samples. As this process repeats over multiple iterations, each new tree increasingly targets the errors left by its predecessors, thereby progressively reducing overall errors and enhancing the accuracy of the model. This method effectively learns from misclassified samples, making the ensemble stronger with each step.

Support Vector Machine (SVM). SVM is another supervised machine learning algorithm widely used for classification and regression tasks. Figure 2.5 illustrates its basic idea by

Fig. 2.5 An example of an SVM

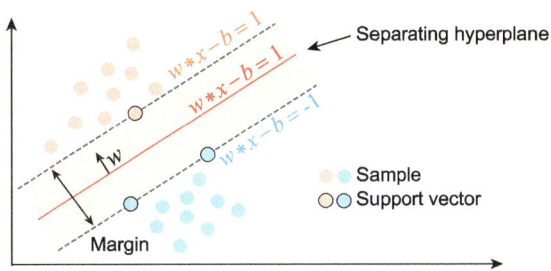

using classification as an example. In this example, SVM aims to find an optimal hyperplane in a high-dimensional space that effectively separates different classes of data samples. The hyperplane is chosen in such a way that the margin between the hyperplane and the nearest data samples of each class is maximized. The data samples closest to the hyperplane, called support vectors (◯◯), play a crucial role in determining the hyperplane. SVM is known for its ability to handle both linearly separable and nonlinearly separable data through the use of different kernel functions. Kernel functions enable the algorithm to implicitly map the input data into higher dimensional spaces, where it becomes easier to find a separating hyperplane. SVM has gained popularity due to its robustness, effectiveness, and generalizability. It has been widely used in a variety of applications such as image classification, text classification, and bioinformatics.

2.2.2   Deep Models

Deep models employ artificial neural networks to learn from data and make decisions. These models consist of computational units known as neurons, structured into layers that form a network. Each neuron processes input data and passes its output to subsequent layers, facilitating a complex hierarchical learning process. This structure allows deep models to learn intricate patterns and relationships within large volumes of data. This makes them particularly effective for tasks such as image recognition, natural language processing, and predictive analytics. Typical examples of deep models include MLPs, CNNs, RNNs, GNNs, DGMs, and transformers.

Multi-Layer Perceptron (MLP). As shown in Fig. 2.6, an MLP contains an input layer, several hidden layers, and an output layer. The neurons between consecutive layers are fully connected, which indicates that the inputs of a neuron are the outputs of all neurons in the previous layer. The output of a neuron is calculated as $g(\sum_i w_i x_i + b)$, where $\sum_i w_i x_i$ is the weighted sum of its inputs, b is the bias, and g is the activation function. The weights w_i and the bias b are learned during training. Typical activation functions are nonlinear, including hyperbolic tangent $g(v) = \tanh(v)$, sigmoid $g(v) = 1/(1 + \exp(-v))$, and rectified linear unit (ReLU) $g(v) = \max(0, v)$.

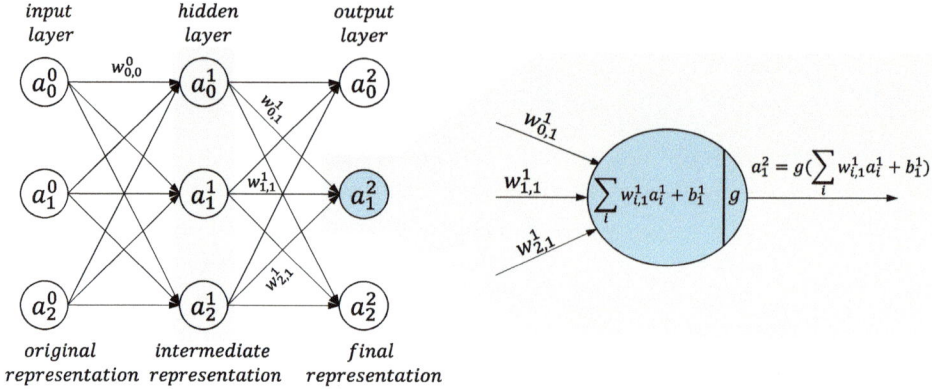

Fig. 2.6 A typical example of an MLP

Convolutional Neural Network (CNN). A CNN employs convolutional operations instead of standard matrix multiplication in one or more of its layers [133]. The convolution makes it suitable for efficiently processing multi-dimensional array data. Therefore, CNNs are widely used in computer vision tasks, including image classification, object detection, and image segmentation. As shown in Fig. 2.7a, a CNN usually consists of three types of layers: convolutional layers, pooling layers, and fully connected layers. Figure 2.7b illustrates the working mechanism of these three layers. The convolutional layers use a set of learnable kernels to detect specific patterns in the input. Each kernel computes the dot product with a region of the input and records the result at the corresponding position on the feature map. By repeating this process for every region of the input, the feature map captures the presence and intensity of a specific pattern across all positions. The pooling layers then perform downsampling to reduce the spatial dimensions of the feature maps, which consequently decreases the number of parameters in subsequent layers. For example, a 2x2 max pooling layer halves the size of feature maps by selecting the maximum value from each 2x2 region of the input. Finally, the fully connected layers produce the class predictions based on the features extracted by the previous layers.

Recurrent Neural Network (RNN). RNNs are particularly designed to process sequential data. They employ recurrent connections for this purpose. These connections enable information to be carried over and propagated through different tokens of a sequence. As shown in Fig. 2.8, an RNN maintains hidden states (h_t) for individual input tokens (x_t) and updates these states as the sequence unfolds: $h_t = V h_{t-1} + U x_t$. The output of each token can then be calculated by $o_t = W h_t$. However, vanilla RNNs, a basic form of RNNs without advanced gating mechanisms, often encounter issues with recursive gradient updates such as the vanishing gradient problem. To address these issues, various gated units, such as gated recurrent units and long short-term memory networks, have been developed to improve recurrent computations. These units control the flow of information and enable the network to selectively maintain or forget information. This capability is crucial for effectively han-

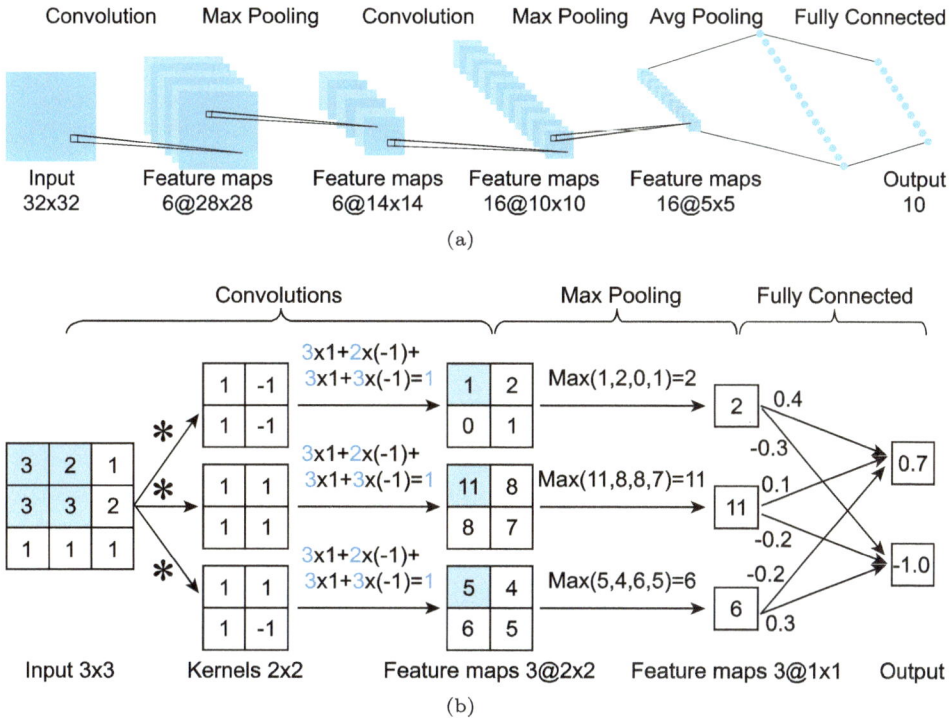

Fig. 2.7 An example of a CNN: **a** the typical structure of a CNN; **b** the working mechanism of convolution layers, max pooling layers, and fully connected layers

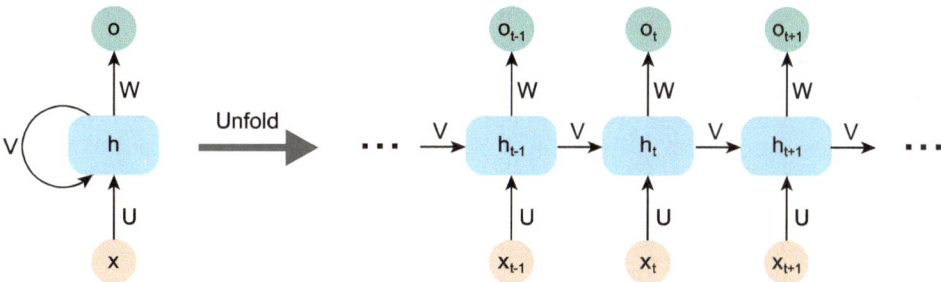

Fig. 2.8 A typical example of an RNN

dling sequences with long-term dependencies and improving the performance of recurrent computations.

Graph Neural Network (GNN). GNNs, designed for graph data analysis, tackle the challenge of effectively learning from both graph nodes and edges. These networks efficiently capture the complex relationships and interdependencies between nodes in a graph. Typically, they employ a three-step method [269]. First, for each edge in the graph, a message is

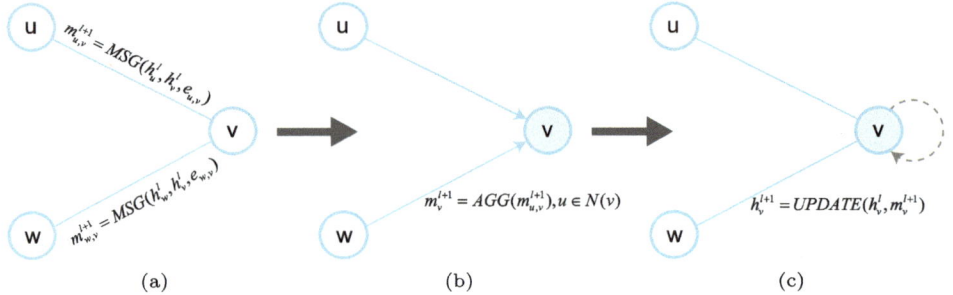

Fig. 2.9 The working mechanism of a GNN: **a** create message; **b** aggregate information; **c** update hidden state

created to include information from the associated nodes and their relationship (Fig. 2.9a). Second, each node aggregates information from all its neighbors (Fig. 2.9b). This allows the model to learn representations for each node based on both its own features and the characteristics of its neighbors. Third, the hidden state of each node is updated using the aggregated information from its neighbors and its previous state (Fig. 2.9c). This iterative process, often involving multiple layers, enables GNNs to capture the hierarchical patterns presented in graph data. As a result, GNNs excel in tasks such as social network analysis, molecular structure interpretation, and recommendation systems, where understanding the relationships and interconnections is crucial. Different ways of implementing the above three steps lead to different variants of GNNs, such as graph convolutional networks, graph attention networks, and graph isomorphism networks.

Deep Generative Model (DGM). DGMs aim to understand and replicate the distribution of training data. Their goal is to produce data samples that not only follow the distribution but also share the characteristics of the training data. By leveraging the power of deep neural networks (DNNs), complicated DGMs can generate data samples that are both realistic and diverse. Consequently, they have been adopted across a wide range of applications such as image synthesis, text generation, music composition, and data augmentation. There are multiple types of DGMs, but the most popular are generative adversarial networks (GANs) and variational autoencoders (VAEs).

As shown in Fig. 2.10a, a GAN model consists of two sub-networks: a generator and a discriminator. GAN uses a competitive process in which the generator aims to create synthetic data samples $G(z)$ indistinguishable from real data samples, from a random latent vector z. Meanwhile, the discriminator seeks to accurately differentiate these synthetic samples $G(z)$ from real data samples x. This interaction forms a competitive process where the discriminator strives to identify samples generated by the generator, and the generator continuously improves to produce increasingly realistic samples to fool the discriminator. Through this process, both sub-networks enhance their capabilities, which in turn leads to more realistic synthetic samples.

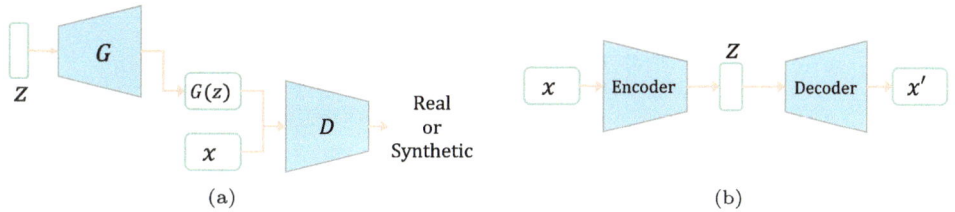

Fig. 2.10 The architecture of two popular DGMs: **a** GAN; **b** VAE

Unlike a GAN model that learns to fool the discriminator, a VAE model (Fig. 2.10b) learns to reconstruct the input samples. The VAE architecture employs an encoder to map the high-dimensional input sample x to the low-dimensional latent vector z. Subsequently, a decoder is used to reconstruct the sample x' from its low-dimensional latent vector. This process focuses on capturing the essential features of the input data in the latent space, and then enables the VAE to efficiently reconstruct the original data with a focus on preserving its key characteristics. The learning goal of VAEs is to optimize the latent space to ensure that the reconstructed samples closely resemble the original input.

Transformers. Unlike RNNs that process data sequentially or CNNs that emphasize local features, transformers learn to capture the relationships among all data tokens (e.g., words of a sentence) simultaneously using self-attention mechanisms. A transformer model consists of two key components, the self-attention-based encoder and decoder (Fig. 2.11a). The transformer encoder takes a sequence of tokens as input and processes them through multiple stacked self-attention blocks. Each block contains multiple attention heads (Fig. 2.11b), and each head understands the relationships among all tokens in the input sequence through scaled dot-product attention. The transformer decoder works similarly to the encoder. In addition to self-attention, it incorporates encoder-decoder attention, which enables it to focus on the most relevant tokens of the output of the encoder.

Transformers have achieved great success in various natural language processing tasks such as machine translation, text summarization, and sentiment analysis. Recently, their application has expanded to computer vision tasks, such as ViT [60], where they have achieved performance comparable to and even better than CNN. Transformers are now the fundamental building blocks for many large-scale pre-trained models such as GPT [25, 187, 188] and CLIP [200].

2.2.3   Foundation Models

A foundation model is a large-scale machine learning model pre-trained on extensive cross-domain data, typically through self-supervised learning methods [22, 268]. These models typically feature hundreds of millions to billions, or even trillions, of parameters. The sheer scale of parameters and the extensive training data enable these models to capture general

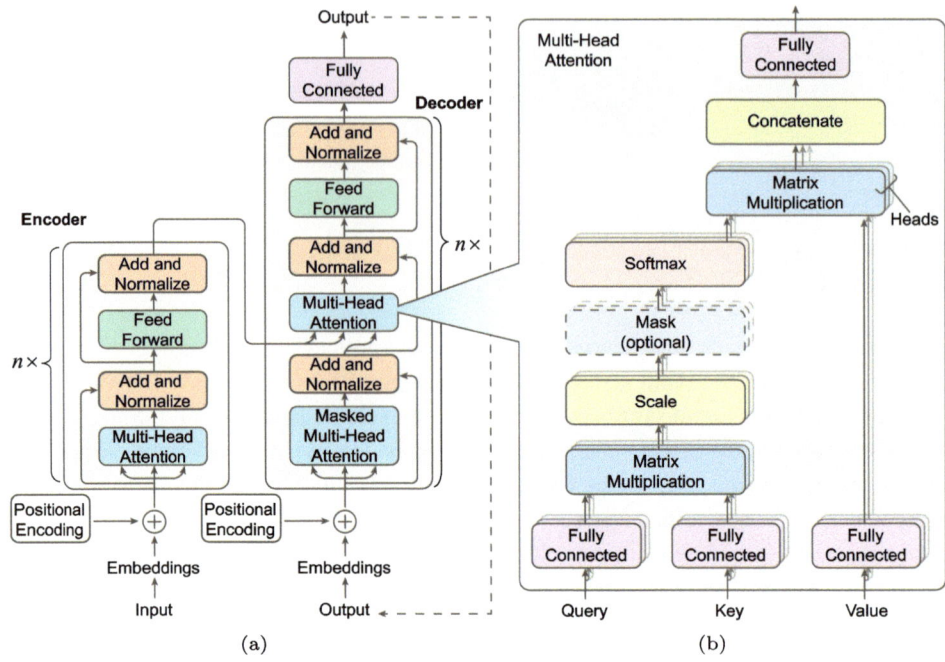

Fig. 2.11 The architecture of transformer and the working mechanism of multi-head attention blocks: **a** the architecture of transformer with an encoder and a decoder; **b** the working mechanism of multi-head attention. This is inspired by Vaswani's work [236]

knowledge about the world. They serve as a "foundation" to effectively adapt to a variety of downstream tasks such as natural language understanding, image recognition, question answering, and image segmentation [22]. Notable examples of such models include BERT [54], ViT [60], InternImage [247], CLIP [200], and GPT series models [25, 187, 188].

Bidirectional Encoder Representations from Transformers (BERT). BERT is one of the most famous foundation models in the field of natural language processing. Unlike traditional models that process words in a sentence one by one, BERT employs the encoder part of the transformer architecture, which is naturally suitable to simultaneously analyze both preceding and succeeding context at every layer (Fig. 2.12a). This capability enriches its contextual understanding. BERT is pre-trained on massive textual data through two tasks: (1) masked language modeling that predicts randomly masked words based on their context, and (2) next sentence prediction that predicts whether two given sentences come sequentially in a text. After this pre-training phase, BERT is ready for fine-tuning across a range of NLP applications, including question answering, sentiment analysis, and named entity recognition.

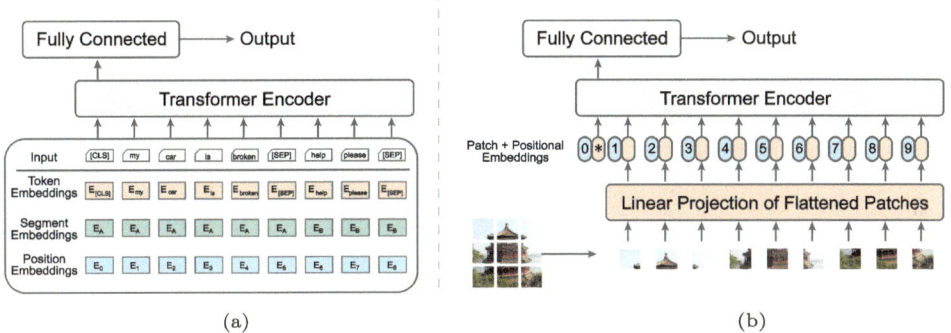

Fig. 2.12 The basic architecture of **a** BERT and **b** ViT

Vision Transformer (ViT). Inspired by the success of transformer architectures in natural language processing, researchers have extended this architecture to develop ViTs for computer vision tasks. Different from traditional CNNs that rely on local convolutions for image analysis, ViTs divide an image into fixed-size patches and transform them into linear embeddings. These embeddings are then processed by the encoder component of the transformer (Fig. 2.12b). This method enables the model to capture long-range dependencies across different regions of an image. Due to its pre-training on vast amounts of images using self-supervised learning, ViTs have achieved state-of-the-art performance on a variety of computer vision tasks.

InternImage. While ViTs have achieved great success in the computer vision domain, researchers continue to explore large-scale foundation models based on CNNs. One of the most notable examples of such exploration is InternImage. Unlike traditional CNNs that utilize dense kernels, InternImage employs dynamic sparse kernels. The benefits of using dynamic sparse kernels are three-fold. First, they facilitate the capture of long-range dependence between pixels, a crucial feature for tasks such as object detection and instance segmentation. Second, these kernels support adaptive spatial aggregation. By dynamically weighting different regions, the model learns more effectively and robustly from a large dataset. Third, the use of dynamic sparse kernels contributes to both computational and memory efficiency. This efficiency allows InternImage to scale effectively in terms of parameter size and training data volume.

Contrastive Language-Image Pertaining (CLIP). CLIP represents a notable advancement in multi-modal learning. In contrast to traditional models that specialize in images or texts, CLIP benefits from training on extensive datasets comprising both modalities. As shown in Fig. 2.13, its architecture includes two main components: a text encoder and an image encoder. These encoders map text and images into a shared latent space. CLIP uses contrastive learning during its training phase to ensure that an image and its corresponding text are close to each other in this latent space. It achieves this by maximizing the agreement between corresponding image-text pairs (diagonal cells in Fig. 2.13) while minimizing

Fig. 2.13 The CLIP model encodes images and their corresponding texts separately through two transformer models and learns to maximize their correspondence in the latent space through a contrastive loss

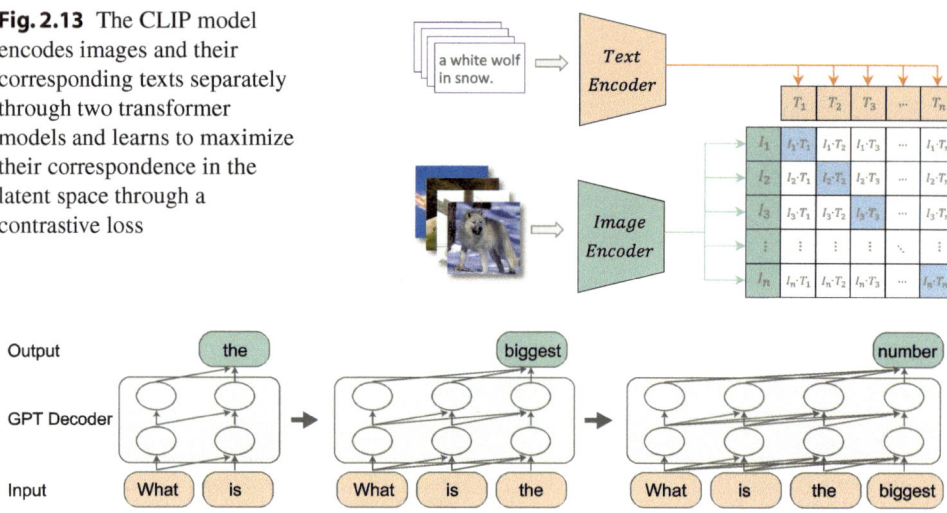

Fig. 2.14 The text generation process of a GPT model

agreement between non-matching pairs (non-diagonal cells in Fig. 2.13). Building on its comprehensive understanding of image-text relationships, CLIP demonstrates exceptional knowledge transfer capabilities. This capability allows CLIP to effectively tackle previously unseen tasks, such as image classification, object detection, and even more complex challenges, through zero-shot and few-shot learning.

Generative Pre-trained Transformer (GPT)*.* GPT is a large language model that leverages the power of the transformer architecture for text generation. As shown in Fig. 2.14, it only employs the decoder part of the transformer architecture because this part is naturally suited for generating the relevant output based on the received input. Training on extensive textual data to predict the next word based on previous words, GPT models are able to capture intricate patterns, grammar, and contextual relationships in language. With their deep and complex architecture, GPT models are effective in generating coherent and contextually appropriate texts. This capability makes them highly effective for applications such as text completion, summarization, and creative writing. Furthermore, the adaptability of GPT models allows for further fine-tuning for specific tasks, which enhances their capabilities in domain-specific language understanding. With these capabilities, GPT has significantly impacted natural language processing. It powers chatbots (e.g., ChatGPT), assists in content generation, and serves as text-based virtual assistants, which have revolutionized the field.

Table 2.1 Relations between data and models. Each cell contains the number of papers

	Tabular	Sequential	MD Array	Graph	Multi-Modal
Regression	2,030	2,850	6,920	846	592
Tree Model	3,460	5,250	12,900	1,220	1,080
SVM	2,970	12,800	45,200	3,290	3,110
MLP	3,490	12,400	26,900	4,920	2,600
CNN	6,010	26,800	118,000	9,450	8,070
RNN	3,010	31,700	24,900	5,710	3,500
GNN	977	2,590	4,040	10,500	859
DGM	936	3,000	7,010	1,100	785
Transformer	3,740	9,260	17,900	4,310	2,670
BERT	3,500	7,100	9,820	3,620	2,430
ViT	690	1,580	7,510	590	960
InternImage	4	8	39	3	8
CLIP	39	58	195	7	72
GPT	740	1,390	1,990	640	407

2.3   Relationships Between Data and Models

To investigate the compatibility of various data types with different models, we utilize Google Scholar to investigate their co-occurrences. The results are summarized in Table 2.1. Our analysis reveals that CNNs, RNNs, and GNNs are particularly well suited for processing multi-dimensional array data, sequential data, and graph data, respectively. Meanwhile, transformer-based models are popular for handling both multi-dimensional array data and sequential data.

In recent years, an increasing number of studies have highlighted the critical role of high-quality training data in developing robust and accurate ML models [155, 184, 266, 270]. However, preparing high-quality training data is a laborious and time-consuming process. It requires users to thoroughly examine a large number of samples, identify potential quality issues, and then identify effective strategies to mitigate these issues. In this context, visualization emerges as a promising method that offers significant assistance in these tasks. Consequently, many VIS4AI methods have been developed to streamline the process of preparing high-quality training data. Since each data sample includes an instance, an annotation, and a feature vector, current methods are classified into three categories: instance diagnosis, annotation diagnosis, and feature engineering [152, 270]. **Instance diagnosis** aims to identify and address issues at the individual data instance level such as missing values, out-of-distribution (OoD) instances, and inexact instances whose content is not easy to recognize. **Annotation diagnosis** focuses on annotation-level issues, including inaccurate annotations, insufficient annotations, and inexact annotations. **Feature engineering** seeks to improve the performance of ML models by adding critical features or removing redundant features. Within each category, three settings are considered, including inaccurate, insufficient, and inexact. Figure 3.1 shows the relationships between the data, three settings, and different visualization techniques for data preparation. Furthermore, Fig. 3.2 uses concrete examples to illustrate the intersections between the three settings and three categories of the associated VIS4AI methods. In practical applications, tasks within these categories are not performed separately. Instead, they are conducted iteratively and collectively throughout the data preparation process. This ensures a continuous improvement in data quality and thereby leads to better model performance.

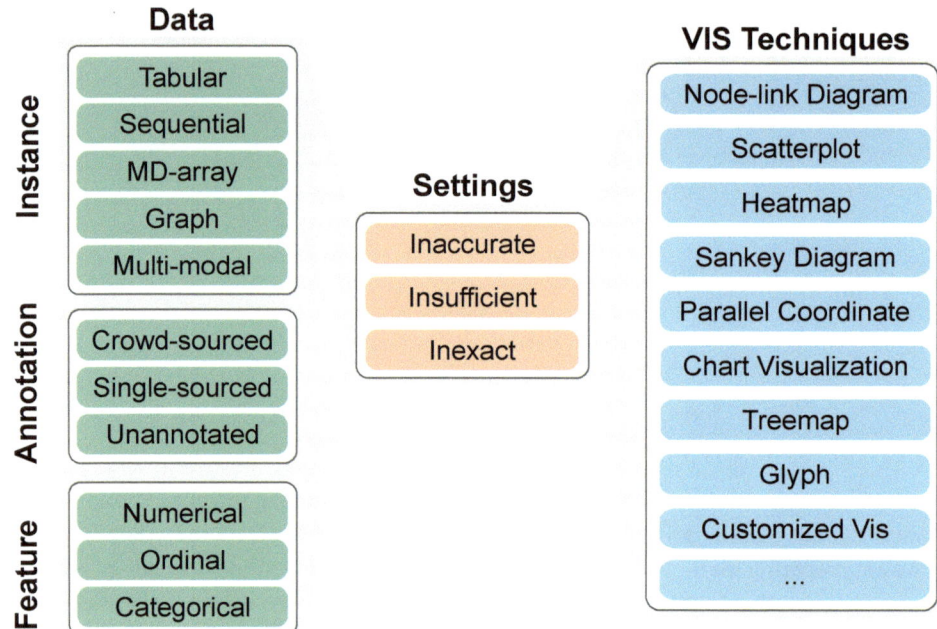

Fig. 3.1 An overview of the categories of data, settings, and visualization techniques for data preparation

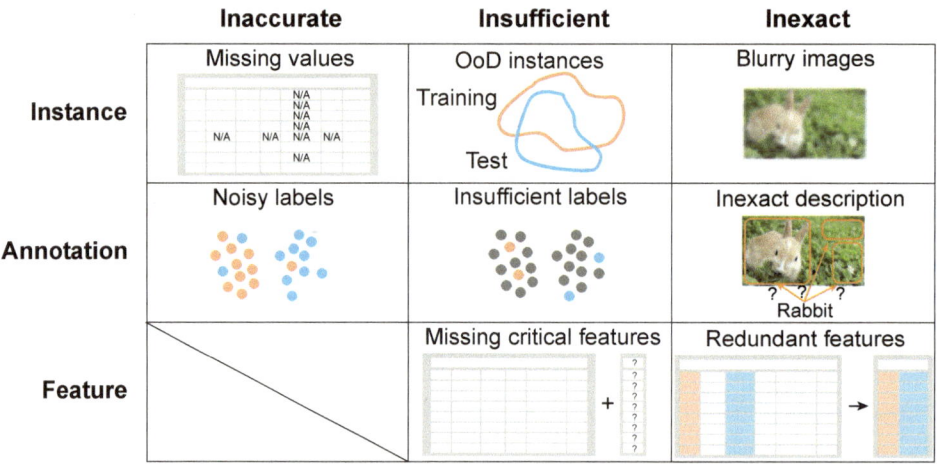

Fig. 3.2 Several representative examples of the intersections between three settings (inaccurate, insufficient, and inexact) and three categories of existing VIS4AI methods (instance diagnosis, annotation diagnosis, and feature engineering)

3.1 Instance Diagnosis

3.1.1 Inaccurate Instances

1	1.3	19	N/A
2	0.9	41	N/A
3	1.3	N/A	N/A
4	1.1	28	77
5	1.4	33	N/A
6	98.4	34	74

Inaccurate instances are those with errors in their attribute values, either due to human mistakes during data entry, sensor errors during data collection, or other external factors. The most common types of inaccurate instances are missing values and abnormal values. Missing values occur when data is not collected or processed correctly, which results in NULL or N/A. An example would be a weight sensor that fails to record data at night. Abnormal values are those that are highly suspect, such as extreme values and abnormal trends. For example, a weight sensor records a value that significantly deviates from the expected range or even records a negative value. Accordingly, we investigate existing VIS4AI studies on handling missing values and abnormal values.

Fernstad and Westberg [72] designed MissiG, a novel visualization to present three missing patterns between attributes, including the amount missingness, the joint missingness, and the conditional missingness. As shown in Fig. 3.3a, each MissiG glyph represents an attribute. The distribution of the values in this attribute is represented by a gray histogram in the left half of the glyph (Fig. 3.3A). The amount missingness represents the instances with missing values in this attribute. It is encoded by the relative height of a light blue block (Fig. 3.3B). Once the user selects an attribute, the joint missingness and conditional missingness will be visualized on all the other attributes and their associated MissiG glyphs. The joint missingness represents instances with missing values in both the selected attribute and the associated attributes. It is encoded by the relative height of a red block (Fig. 3.3C) overlaid on the light blue block. Analyzing joint missingness can help identify the subset of instances with severe missingness across multiple attributes, which may reveal unexpected issues in data collection and preprocessing. The conditional missingness represents the instances with missing values in the selected attribute but not in the associated attributes. It is encoded by a red histogram in Fig. 3.3D, which reveals the distribution of values in the associated attribute for these samples. Analyzing conditional missingness can help understand the reason for missingness and decide better imputation methods to handle the missingness. In addition, MissiG glyphs can be used to enhance existing techniques that are widely used to visualize and analyze missing values such as parallel coordinate plots (PCPs) and heat maps (Fig. 3.3b).

Visplauser [7] is another representative VIS4AI work that enables the interactive examination of data anomalies and improves data quality. It simultaneously supports the analysis of missing values and abnormal values. Once the data is loaded into Visplauser, hundreds of plausibility checks are automatically conducted to evaluate whether there are potential

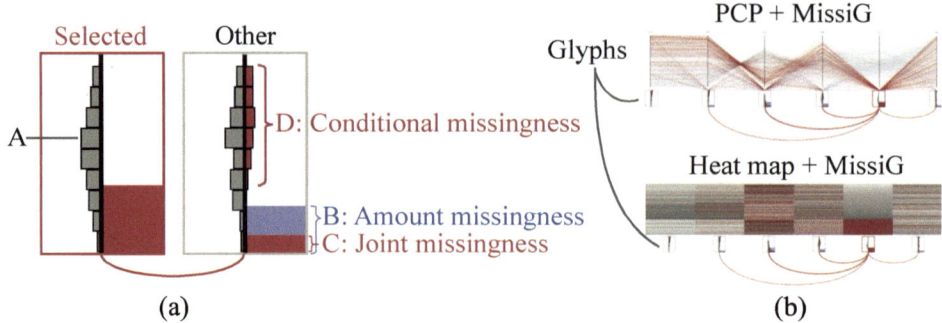

Fig. 3.3 MissiG [72] well presents three missing patterns, including the amount missingness, the joint missingness, and the conditional missingness: **a** MissiG glyphs; **b** integrated with other visualizations

quality issues. These plausibility check results are then organized in a hierarchical tabular layout so that users can examine them at different granularities. In this visualization, each row shows an aggregated plausibility check result of the same anomaly type such as "missing" and "zero at daytime" (Fig. 3.4A). Once the user clicks a row, the detailed check results are displayed using an indented layout. The columns display statistic summaries for further analysis, including the number of checks and data (Fig. 3.4B), the percentage of affected data (Fig. 3.4C), the distribution of affected data over time (Fig. 3.4D), and the severity (Fig. 3.4E). Two linked views, a line chart (Fig. 3.4F) and the original table (Fig. 3.4G), provide additional information about the selected data anomalies. Users can efficiently examine anomalies in detail and generate hypotheses about the possible causes of the identified quality issues during exploration.

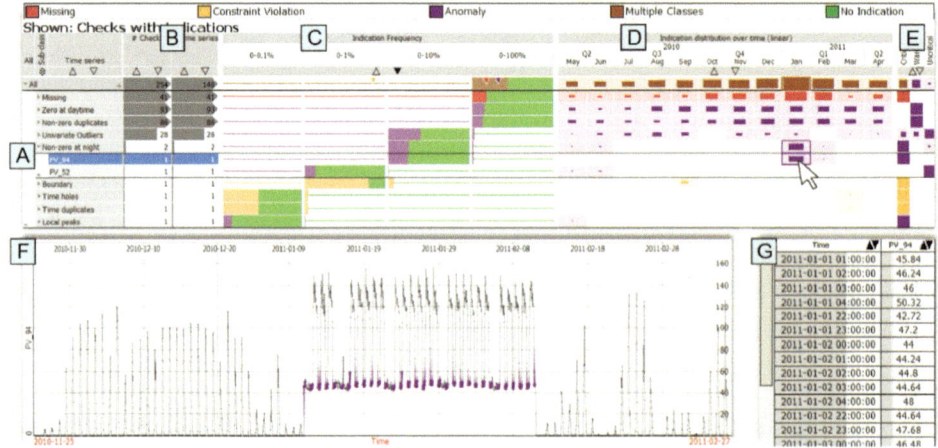

Fig. 3.4 Visplause [7]: a visual analytics tool designed to examine inaccurate instances

There are more VIS4AI studies that empower users to visually analyze and handle missing values and abnormal values. Among them, PCPs and heat maps are the most popular ways to visualize missing values in tabular data due to their intuitiveness and compact form [5, 15, 71]. The strategy of integrating data mining techniques to identify abnormal values and then summarizing them for further analysis is also widely adopted in many visual analytics tools [83, 84, 116].

3.1.2 Insufficient Instances

Insufficient instances occur when the amount of data available for training an ML model is inadequate to capture the complexity and diversity of the targeted problem. For example, consider a cat-dog classifier trained on a dataset that contains only images of dark-colored dogs and light-colored cats. This classifier may rely on color as the key distinguishing factor to make predictions and thus fail to classify dogs or cats of different colors correctly. In such cases, the dogs or cats of different colors are considered as OoD instances, which are not adequately covered by the training instances. VIS4AI methods empower users in identifying these OoD instances and then addressing these issues by collecting additional instances or augmenting existing ones.

OoDAnalyzer [38] is one of the representative methods for detecting and analyzing OoD instances. Inspired by deep ensembles [131], this method employs the ensemble method to detect OoD instances. The OoD score is calculated as the entropy of the predictive distribution among different deep models. Instances with highly inconsistent predictive distribution are more likely to be OoD instances and warrant further examination. Despite the effectiveness of this OoD detection method, it falls short of providing a comprehensive explanation for the detected OoD instances. To address this issue, a similarity-preserving grid visualization is developed to assist users in analyzing OoD instances. By placing similar instances close to each other, users can analyze and understand the root cause of the detected OoD instances in context. The generation of grid visualization consists of two steps. First, the instances are projected as a set of 2D scattered points using t-distributed stochastic neighbor embedding (t-SNE), which preserves their similarities. Second, these 2D points are assigned to the grid cells in a one-to-one manner that guarantees minimal distortion. This is achieved by solving a linear assignment problem between the scatter points and the grid cells. In this grid visualization (Fig. 3.5a), the circles represent training instances, and the squares represent test instances. The color hue encodes the categories of the instances, and the color saturation encodes their OoD scores. OoDAnalyzer offers two modes to facilitate the analysis of OoD instances. In the juxtaposition mode (Fig. 3.5b), the grid visualizations

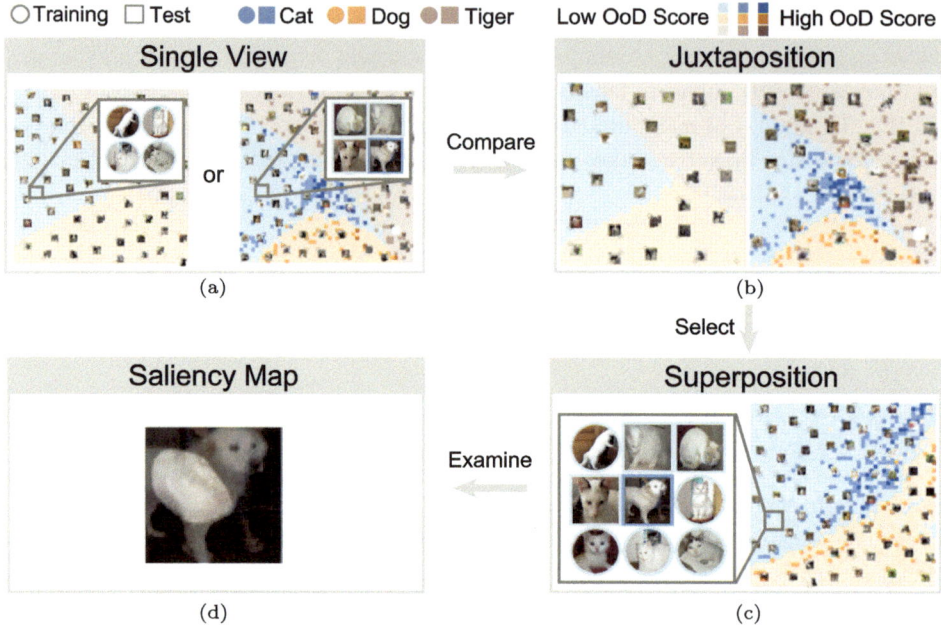

Fig. 3.5 A typical analysis workflow of OoDAnalyzer [38]: **a** examine the distribution of training or test instances; **b** compare training and test instances side-by-side; **c** place the training and test instances together for comparison and zoom into a region of interest; **d** analyze the saliency map

of the training and test instances are displayed side-by-side, which allows users to identify which categories contain more OoD instances. Once the categories of interest are selected, users can switch to the superposition mode (Fig. 3.5c) to directly compare the OoD instances with their relevant training instances. They can further zoom into a local region to examine more instances. In addition, users can analyze the saliency maps of selected instances to identify the reasons for the occurrence of OoD instances (Fig. 3.5d). After identifying the main reasons, they expand the training data to cover these OoD instances, and the model performance is improved.

To alleviate the effort in expanding training data, Gou et al. [82] proposed VATLD, a visual analytics tool to improve the accuracy and robustness of traffic light detectors by generating additional instances that are not well covered by current training instances. It adapts a disentangled representation learning method, $\beta-$VAE [94], to extract semantic representations from the training instances. The extracted semantic representations not only help users understand the bottleneck of model performance but also enable the generation of meaningful instances. A grid visualization (Fig. 3.6a) is employed to display the image content or performance metrics, where the x-axis and the y-axis encode two selected dimensions of semantic representations. Each instance is located in the corresponding cell based on its semantic representation. Since multiple instances may be located in the same cell,

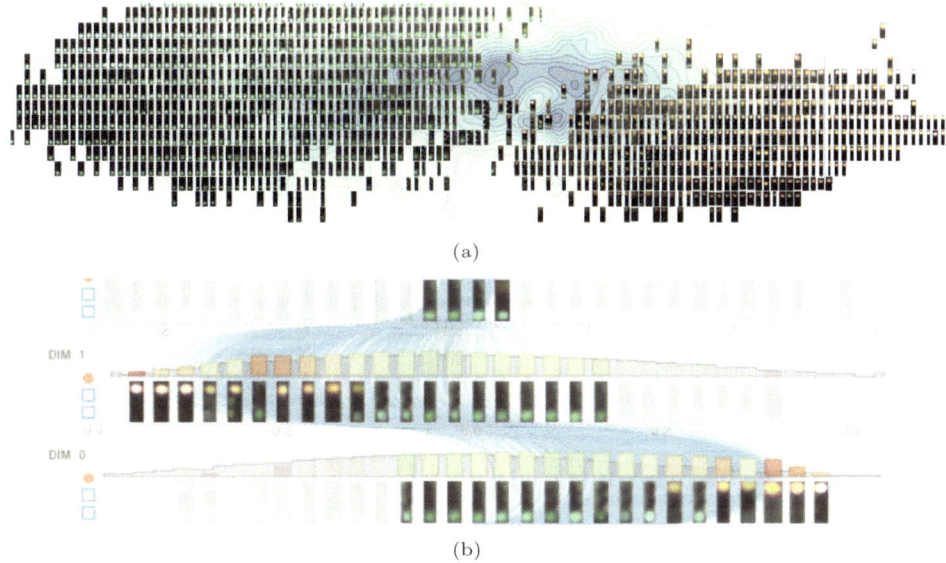

(a)

DIM 1

DIM 0

(b)

Fig. 3.6 The major visualization component in VATLD [82]: **a** a grid visualization to display overall data distribution; **b** a hierarchical parallel coordinate plot to display the semantic representation space

the cell will display the image content of the representative instance or use the fill color to encode the average metrics. To facilitate the exploration of a large number of latent dimensions, these dimensions are hierarchically clustered with an agglomerative clustering method. Alternatively, VATLD employs a hierarchical parallel coordinate plot (hPCP) to facilitate the exploration of the semantic representation space (Fig. 3.6b). The hPCP and the grid layout are coordinated to facilitate the analysis of the training samples. For example, users can select the dimensions of interest from hPCP as the x-axis or y-axis, generate the corresponding grid layout, and highlight the corresponding cells in the grid layout by hovering or brushing on the bins in the hPCP. Additionally, users can employ a lasso to select cells in the grid layout and highlight the corresponding lines and bins in the hPCP. After identifying the failure cases and understanding their semantics, more instances with similar semantics can be generated by data augmentation. The augmented training samples can then improve the accuracy and robustness of the model.

Despite VATLD, there are more VIS4AI studies that augment training instances using deep generative models. For example, He et al. [92] applied VAEs to generate training samples in different driving scenes, which improves the performance of semantic segmentation in autonomous driving. Kwon et al. [130] applied GANs to debias image classification models by generating new data samples that incorporate identified bias factors, such as colors.

3.1.3 Inexact Instances

Inexact instances, such as blurry images or those with poor lighting, pose challenges to machine learning models by making content recognition difficult. Several ML techniques have been developed to improve the quality of such data, including denoising [136], deraining [140], haze removal [90], low-light image enhancement [284], and image restoration [235]. However, these automatic techniques lack support for interactive refinement, which often results in suboptimal outcomes without user intervention to verify the quality of the refined instances. VIS4AI methods integrate these techniques into visualization to facilitate interactive quality improvement of inexact instances.

Roels et al. [207] developed an interactive tool to support electron microscopy image denoising. Users start the process by selecting a specific image and identifying a region of interest within it. The noise level in this region is then automatically estimated so that the initial parameters of the denoising algorithms can be determined. After applying the denoising algorithm with the initial parameters, the original and denoised versions of the region of interest are presented side-by-side for a detailed examination. Users can also change the denoising algorithms or adjust their parameters to see if the denoising results can be further improved. Once the algorithm and its optimal parameters are determined, all the images in the dataset will be processed. These denoised images are used to improve the performance of downstream tasks such as biological object segmentation.

Several other VIS4AI studies also demonstrate that interactive instance quality improvement benefits from human feedback. For example, Weber et al. [253] proposed a human-in-the-loop framework to interactively restore damaged images. This system enables users to modify the mask of damaged regions, which are then sent back to the restoration model to generate better restoration results. Similarly, Fischer et al. [74] developed NICER, a visual analytics tool that allows users to directly adjust the algorithm parameters, effectively steering the image enhancement process to generate the desired images needed for their applications.

3.2   Annotation Diagnosis

3.2.1   Inaccurate Annotations

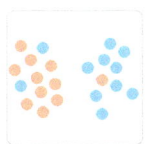

Inaccurate annotations, such as noisy labels and imprecise bounding boxes, will hamper the performance of ML models. Many VIS4AI methods have been proposed to facilitate the identification and correction of inaccurate annotations. These methods can be classified into two categories: improving datasets with crowd-sourced information and improving datasets without crowd-sourced information.

LabelInspect [157] is one of the representative VIS4AI methods to improve annotations in datasets with crowd-sourced information. It employs a max-margin-majority-voting (M^3V)-based crowd-sourced model [148] to infer the annotations from noisy crowd-sourced annotations and incrementally updates the annotations based on expert validation. LabelInspect facilitates the interactive validation of crowd-sourced annotations by offering three coordinated visualizations: a confusion visualization, a sample visualization, and a worker visualization. The confusion visualization (Fig. 3.7A) shows whether workers confuse two or more classes using the confusion matrix. Once users select the classes of interest, the

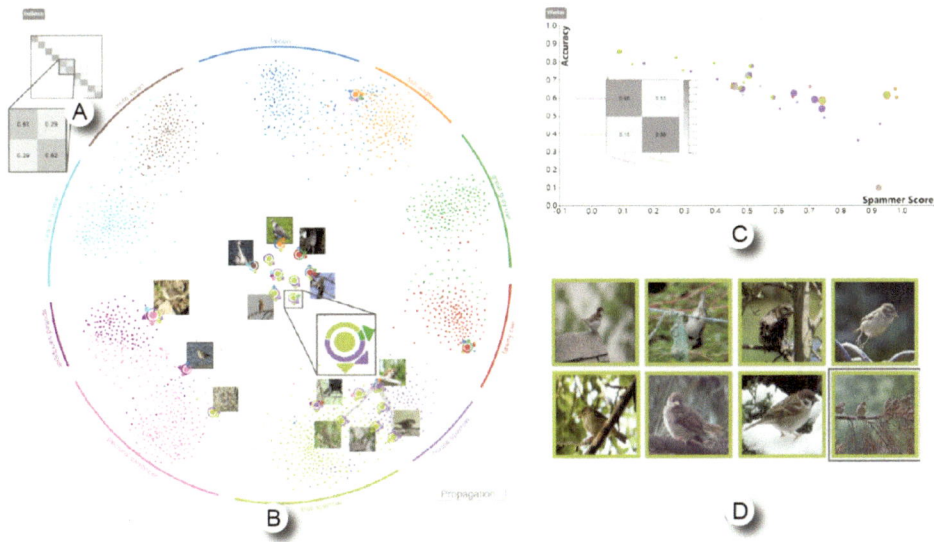

Fig. 3.7 LabelInspect [157]: a visual analytics tool designed to improve the quality of crowd-sourced annotations through coordinated visualizations

corresponding samples will be displayed in the sample visualization (Fig. 3.7B). This visualization is designed as a circular-based constraint layout. The outer arcs represent classes,
where the length of each arc encodes the number of samples in that class. The corresponding
samples are plotted within the circle, where the recommended uncertain samples are encoded
using glyphs, and the remaining samples are encoded using simple dots. To facilitate the
analysis of samples, their positions are determined by a constrained t-SNE projection, which
not only places similar samples together but also pulls certain samples toward their inferred
class and keeps the stability after the model is updated. The worker visualization (Fig. 3.7C)
illustrates worker reliability and enables the validation of crowd-sourced annotations from
worker-level data. Workers are visualized in a scatterplot where the x-axis is the spammer
score, and the y-axis is the accuracy according to inferred annotations. Each dot in the scatterplot represents a worker, and the dot size encodes the number of annotated samples. A
confusion matrix can be generated to examine the detailed behavior of any selected worker.
These three coordinated visualizations enable users to iteratively explore and cross-validate
sample annotations alongside worker reliability. For example, when users hover over a sample, the workers who label this sample will be highlighted. When users hover over a cell of
a worker's confusion matrix, the corresponding samples will be highlighted in the sample
visualization. Their images are also displayed using a grid visualization (Fig. 3.7D). Validation on either side will update the uncertain samples and unreliable workers, and the
propagation of analysts' validations will also update both inferred annotations and workers'
spammer scores.

   Despite its effectiveness in improving crowd-sourced annotations, LabelInspect is not
suitable for datasets lacking such information. For these datasets, Xiang et al. [260] developed DataDebugger to support the examination and correction of their annotations. To
efficiently examine large-scale annotations, a hierarchical scatterplot is developed. As illustrated in Fig. 3.8, the hierarchy is constructed in a bottom-up manner. Let L_0 be the bottom
layer that contains all samples in the dataset. Level $L_k (k > 0)$ is constructed by sampling
25% of the samples from the previous layer L_{k-1}. In the meantime, each sample in L_{k-1} is

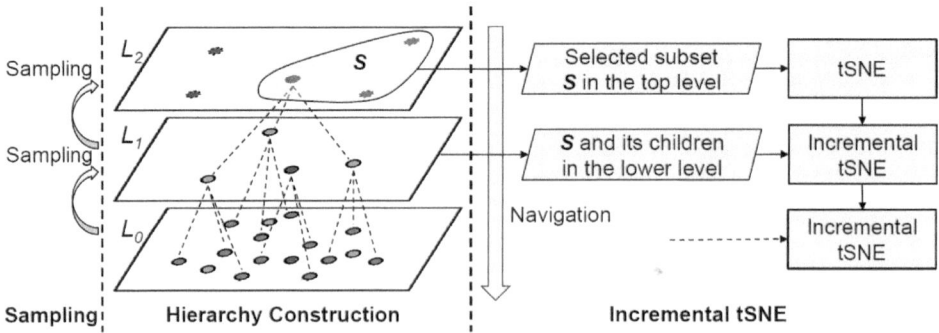

Fig. 3.8 The hierarchical scatterplot proposed by DataDebugger [260]

assigned as the child of its nearest sample in L_k. This process repeats until the number of samples in the top layer is less than a certain threshold. During the sampling process, both the overall distribution and outliers should be preserved to facilitate the identification of incorrectly annotated samples. To achieve this, an outlier-biased density-based sampling is developed. The sampling probability of i-th sample is set as $p_i = 1/\rho_i + \pi_i$, where ρ_i is the local density and π_i is the outlier probability. With the constructed hierarchy, an incremental t-SNE is employed to ensure a smooth transition between different levels. Specifically, once the user selects a set of samples on the t-SNE of level L_k and drills down to display the relevant samples in L_{k-1}, a randomly sampled subset of the selected samples at level L_k are considered as virtual anchor samples during the layout process. The incremental t-SNE maintains the distances between the selected samples at level L_k and virtual anchor samples to provide a more stable projection during the top-down navigation. Once users identify incorrectly annotated samples and make corrections, an automatic annotation correction algorithm is employed to propagate users' annotations to the entire dataset, which significantly reduces the labor required in the correction process.

In addition to LabelInspect and DataDebugger, there are several similar VIS4AI methods designed to improve the quality of annotations in datasets with crowd-sourced information [193, 194] or without crowd-sourced information [3, 13, 190, 267]. For example, Park et al. [193] developed C^2A that simultaneously visualizes crowd-sourced annotations and worker behavior to help doctors identify malignant tumors in clinical videos. Paiva et al. [190] combined multi-dimensional projections and neighbor-joining tree visualization to better explore misclassified samples, which are more likely to be incorrectly annotated and require detailed examination. Yang et al. [267] developed Reweighter, an interactive reweighting tool to mitigate the negative effects of low-quality annotations by reducing their weights.

3.2.2 Insufficient Annotations

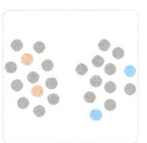

Another challenge in developing a well-performing model is insufficient annotations. The most straightforward solution is to collect more annotations. In this vein, several VIS4AI methods have been developed to improve the efficiency of the annotation process. Furthermore, recent advances in semi-supervised learning have led to the development of VIS4AI methods that incorporate these techniques to better utilize unannotated instances.

Improving annotation efficiency. An effective way to improve annotation efficiency is to place similar instances close to each other. This allows users to analyze multiple instances in context and annotate similar instances together. Moehrmann et al. [176] developed a labeling tool based on the self-organizing map (SOM). The SOM is a dimensionality reduction method that projects high-dimensional data, such as images, into low-dimensional grid cells while preserving similarity relationships. In this SOM-based visualization, the image content is displayed in the projected cells for examination. If multiple images are projected into the same cell, a representative image will be displayed in each cell. Users can then easily label images by selecting an individual cell or multiple cells with a rectangle or a lasso. Once labeled, all the images within the selected cells are assigned the same label. However, this method might introduce labeling errors since the SOM can sometimes place images of different classes into the same cell. To address this issue, the tool enables users to zoom into specific regions, examine individual images, and correct the identified labeling errors.

In addition to placing similar instances close to each other, another effective way is to filter the instances belonging to the same class and then annotate them collectively. Hoque et al. [103] proposed visual concept programming, which enables the interactive construction of labeling functions, which filter images of interest and assign labels to them. Specifically, the process begins with the extraction of visual concepts such as transport, rail, and water. This is achieved by segmenting images into different segments, clustering similar segments into visual concepts, and then tagging concepts with descriptive natural language terms. Users can explore these concepts in the projection view (Fig. 3.9A) and the rank view (Fig. 3.9B). These two views are coordinated to help identify and select a primary concept of interest. Next, the concept feature view (Fig. 3.9C) utilizes a bar chart to display the distribution of segments belonging to the primary concept. When users select a secondary

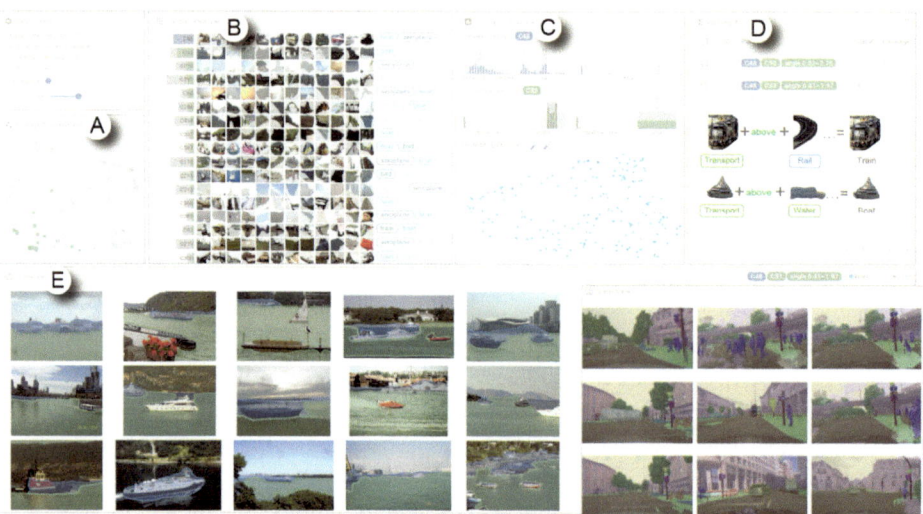

Fig. 3.9 Visual Concept Programming [103]: a visual analytics tool designed to create labeling functions for large-scale datasets

concept, the bar chart highlights the subset of these segments that co-occur with the secondary concept. Users can brush on the bar chart to filter the corresponding segments. Through interactive filtering and examination of related images, the user constructs labeling functions (e.g., transport+rail→train), which are displayed in the labeling rules view (Fig. 3.9D). Once the user selects a labeling function, the corresponding images that satisfy the filtering condition are displayed in the scene view for a thorough examination (Fig. 3.9E). If users are satisfied with the filtering results, the labeling function can be applied to label all the filtered images in a single click, thereby streamlining the annotation process. Otherwise, they can adjust the filtering condition to iteratively refine the labeling function.

The strategy of simultaneously annotating multiple instances has also been widely adopted in many VIS4AI methods to improve annotation efficiency. For example, the integration of dimensionality reduction techniques with grid layouts or scatterplots has been shown to be effective in interactively annotating various data types, including mobile eye-tracking data [127], color strategies used in films [87], and spambots in social media [120]. Interactive filtering is also widely used to annotate instances. For example, MediaTable [210] enables users to interactively filter and sort video segments based on their attributes, which facilitates identifying video segments of the same class. Stein et al. [225] provided a natural language GUI for specifying filtering rules and find patterns of interest in soccer match videos. He et al. [89] extracted human-understandable events from videos and used them to interactively filter videos for annotation.

Utilizing unannotated instances. Instead of annotating more instances, some recent VIS4AI studies handle insufficient annotations by better utilizing unannotated instances [37, 265]. Chen et al. [37] developed DataLinker to help users explore, understand, and improve the annotation propagation process in graph-based semi-supervised learning (GSSL) models. In GSSL, the annotations are propagated to unannotated instances via a graph that describes the similarity relationships between instances. DataLinker employs a river metaphor to provide an overview of the annotation propagation process (Fig. 3.10A). Users can identify problematic instances with frequent annotation changes in this overview. To better understand the impact of instances and their relationships on annotation propagation, a hybrid visualization is developed to simultaneously display the sample distribution and the graph structure (Fig. 3.10B). In this visualization, nodes and edges represent instances and their similarity relationships, respectively. As the number of edges is usually large, they are displayed on demand upon selecting a region of interest. In addition, the edges with the same starting classes and ending classes are grouped together to reduce visual clutter. To effectively explore the edges in this graph, a Voronoi-based space partition and a bar-chart-based distribution visualization are combined to give an overview of the edge distribution. The layout space is partitioned into several regions based on each class. A bar chart is displayed for each class and shows the distributions of homogeneous edges (colored) that connect instances of the same class and heterogeneous edges (gray) that connect instances of different classes. A higher proportion of heterogeneous edges indicates greater confusion of this class with others and warrants further examination.

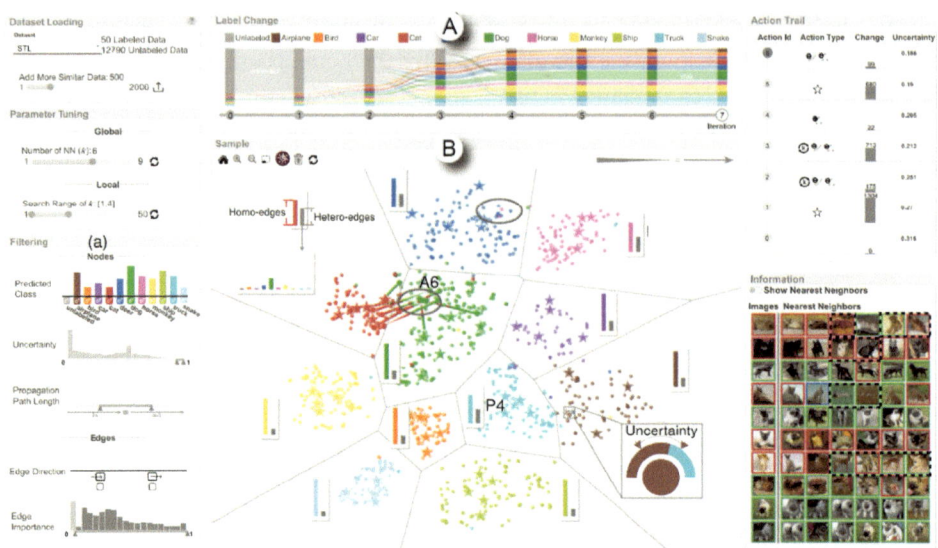

Fig. 3.10 DataLinker [37]: a visual analytics tool designed to explore, understand, and improve the annotation propagation process in graph-based semi-supervised learning models

To better examine the samples of interest and make informed adjustments to the graph structure, DataLinker provides two main interactions: graph filtering and graph construction. Graph filtering seeks to identify the samples of interest. It employs scented widgets [255] to filter out irrelevant nodes and edges, which allows users to focus on those of interest (Fig. 3.10a). Graph construction follows a coarse-to-fine strategy and allows users a make efficient changes to the graph structure at the global, local, and individual levels. At the global level, users can examine the overall sample distribution and identify problematic samples that need label corrections and classes that contain insufficient annotated instances. After adjusting labels and augmenting instances at the global level, users can interactively adjust the k value in the k-NN graph at the local level. Further individual refinements, such as removing incorrect edges and noisy samples, will be carried out for the remaining small number of samples with high uncertainty.

Several other VIS4AI studies also utilize unannotated instances to decide the most valuable instances to annotate, thereby greatly reducing the amount of required annotations. For example, FSLDiagnotor [265] utilizes the sparse subset selection algorithm to recommend the most valuable instances that improve model performance in few-shot learning. Users can analyze the recommendation results, explore their coverage, and annotate these instances to achieve better model performance. A similar idea of exploring and annotating the most valuable instances has been successfully used in various tasks, including text document retrieval [93], sequential data retrieval [137], trajectory classification [113], identifying relevant tweets [222], and argumentation mining [223].

3.2.3   Inexact Annotations

Inexact annotations are those that are correct but lack the desired level of precision. For example, the caption of an image might not exhaustively describe all objects and their locations. For a video clip, the annotations of actions in the video might only indicate their category labels without providing precise boundaries. Early efforts have been made to better utilize inexact annotations with the help of visualization.

Chen et al. [39] developed MutualDetector to better utilize inexact image captions to improve object detection results. The basic idea is that the labels extracted from the captions should be consistent with the labels of detected objects. Thus, the training label extractor and object detector can reinforce each other by adding the consistency constraint. However, due to the noisy nature of captions and unsatisfactory detection results, mismatches between the extracted labels and detected objects often occur, and thus require user validation. MutualDetector utilizes a node-link-based set visualization to support the exploration of matching relationships between the extracted labels and detected objects. It begins by simultaneously clustering the labels and images so that (1) similar labels (images) are clustered together, and (2) labels that are matched with similar images are clustered together, and vice versa. This is formulated as a co-clustering problem. In order to facilitate the exploration of large number of label clusters and image clusters, the co-clustering algorithm is recursively applied to build the label and image hierarchies in a top-down manner. Specifically, it involves fixing the image clusters and dividing each label cluster into sub-clusters using the co-clustering algorithm and vice versa. This process repeats until the number of labels in a cluster is smaller than a certain threshold. With the built hierarchy, a node-link-based set visualization is developed to display label clusters, image clusters, and their relationships. Label clusters are placed on the left side and visualized as an indented tree (Fig. 3.11A), inspired by the Windows File Explorer. Image clusters are placed on the right side as a matrix (Fig. 3.11B) to provide an overview, where each row with ten representative images represents one image cluster. Links between the indented tree and matrix represent the matching relationships between label clusters and image clusters (Fig. 3.11C). A link of a red dashed line indicates the number of mismatches between the respective image cluster and label cluster is larger than a given threshold. Users can analyze the mismatches between labels and images and provide validations. If the labels extracted from captions are incorrect, they can remove the improper words that lead to the wrong extraction. If the detected bounding boxes are imprecise or incorrect, they can directly modify the bounding boxes. If both the extracted labels and the bounding boxes are correct, but the confidence values of the detected objects are low, users can increase the weight of the consistency constraint so that the object detector becomes more confident in detecting these objects with consistent labels.

Fig. 3.11 MutualDetector [39]: a visual analytics tool designed to interactively improve both captions and bounding box annotations

ActLocalizer [40] is another representative work for interactively refining inexact annotations in temporal action localization. It identifies the boundaries and categories of actions in videos with one single annotated frame per action. ActLocalizer begins by training an initial action localizer using these inexact annotations to detect actions. It then displays uncertainty values for each action category in a category list (Fig. 3.12A). When a user selects a category for exploration, the tool hierarchically clusters the actions of this category and displays them in a storyline visualization (Fig. 3.12B). This allows users to explore the detection results of each action and the alignment relationships between similar actions such as common human motions. In this visualization, each contour represents an action cluster, and each line represents an action. Circles (●) on the lines represent the unannotated frames of the action, and the stars (★) represent the annotated frames. The horizontal positions of the frames show their sequential orders, while the vertical distances reflect the alignment relationships between actions. To facilitate the identification of abnormal actions with unusual lengths, ActLocalizer uses bar charts to display the distribution of action lengths in each cluster (Fig. 3.12C). After identifying abnormal actions, the user can correct imprecise boundaries, incorrect category labels, and misalignment relationships between actions. These corrections are then propagated to other actions, which improves the overall quality of the annotations.

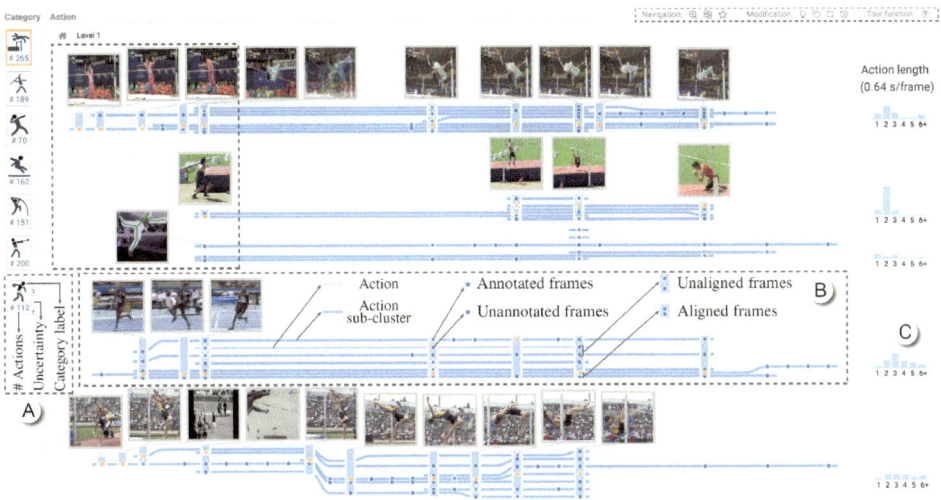

Fig. 3.12 ActLocalizer [40]: a visual analytics tool designed to iteratively improve the performance of action localizers through minimal user corrections on action annotations

3.3   Feature Engineering

3.3.1   Insufficient Features

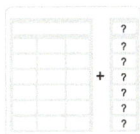

Insufficient features refer to cases where the existing attributes are not yet effective in constructing a high-performing model. In these cases, users seek to create new attributes that are important to the target tasks. This can be achieved by either transforming existing attributes into new ones or integrating additional attributes from external data sources.

Chatzimparmpas et al. [35] developed FeatureEnVi to support interactive feature engineering. In Fig. 3.13A, a beeswarm plot is employed to display the distribution of samples, where the x-axis encodes the predicted probability of each sample. The samples are thus categorized into four groups from left to right, namely "worst," "bad," "good," and "best." This view provides an overview of the current model performance on different samples. To support interactive attribute selection, FeatureEnVi integrates five automatic attribute selection techniques to compute the importance values of different attributes. The importance values are then depicted as a heat map (Fig. 3.13B), where each row represents an attribute, and each column represents a selection technique. Users can then compare different techniques and interactively select attributes that are important to model predictions. After

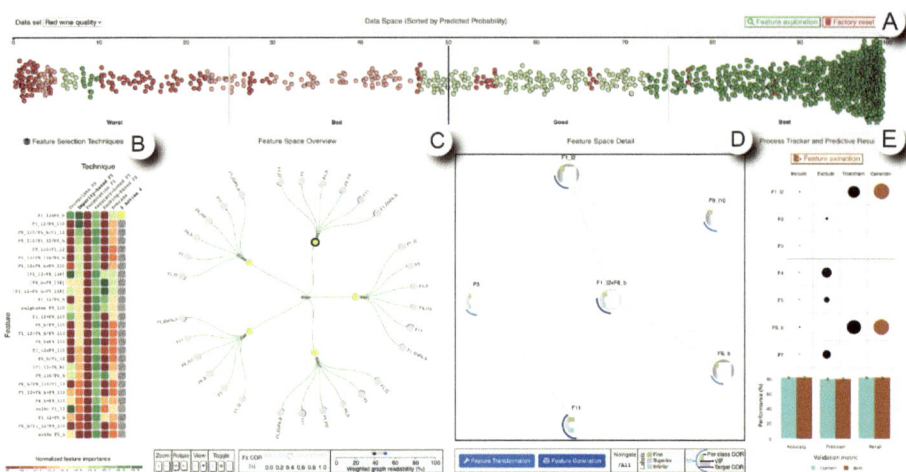

Fig. 3.13 FeatureEnVi [35]: a visual analytics tool for interactive feature engineering

selecting important attributes, a radial tree is utilized to provide an overview of the attributes along with statistical measures such as their mutual information with the ground-truth labels (Fig. 3.13C). In this tree, the five yellow nodes within the tree represent "all" samples and the four groups of samples partitioned in Fig. 3.13A, respectively. The gray nodes outside represent the selected attributes, with statistical measures encoded using glyphs. Upon choosing a specific attribute, a graph visualization is generated to display more statistical measures as well as the correlation between this attribute and other relevant attributes (Fig. 3.13D). Through this view, users can also interactively generate new attributes using mathematical functions such as taking the natural logarithm of one attribute or summing two attributes. All actions, including attribute selection and attribute generation, are logged in a process tracker (Fig. 3.13E), which provides transparency and traceability throughout the feature engineering process.

Cashman et al. [31] developed CAVA to empower developers to create new attributes to augment an existing tabular dataset using knowledge graphs. For example, an existing dataset might contain the attribute "country" but does not contain attributes like "country code" or "average population in the past ten years." After loading this dataset into CAVA, the existing attributes are listed in Fig. 3.14A. Once users select an attribute like "country," the relevant attributes are extracted from the knowledge graph and visualized in another list (Fig. 3.14B). More detailed information, such as the data type, the source of the attributes, and the distribution of values, are displayed on demand (Fig. 3.14C). Once they find a useful attribute, they can click the "+" button in the corresponding row to add it to the current data. However, during this process, the relationships between entities in the existing dataset and entities from the knowledge graph could be complicated. For one-to-one relationships such as "country − country code," the attribute values from the knowledge graph can be

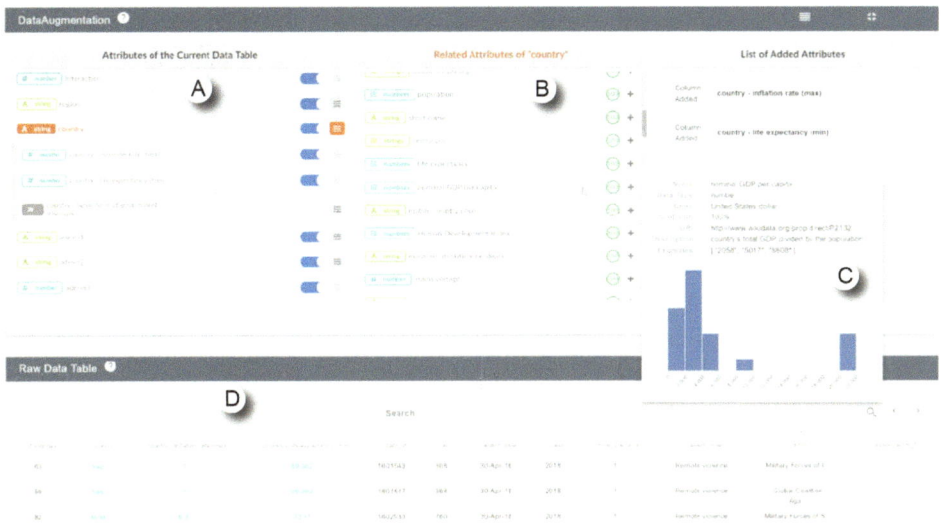

Fig. 3.14 CAVA [31]: a visual analytics tool designed to augment datasets with extra attributes

directly assigned to corresponding entities in the dataset. For one-to-many or many-to-many relationships, users need to specify how the attribute values are aggregated (e.g., max, min, average) before they are added to the dataset. After adding attributes, a preview of the dataset is displayed below for examination (Fig. 3.14D).

3.3.2   Inexact Features

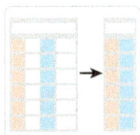

Inexact features refer to redundant attributes that are not helpful and need to be removed for better performance. VIS4AI forms an interactive process to help examine and select the most useful features for model training [36].

Some VIS4AI methods directly visualize attributes for examination and selection. For example, Artur et al. [9] developed RadViz to analyze the attribute space. In RadViz, class labels are uniformly arranged on a unit circle and conceptualized as anchors. Each attribute is connected to all the anchors with springs, where the ideal lengths are determined by the correlations between the attribute and the class labels. The positions of attributes are finalized when the springs reach a state of equilibrium. Through this view, users can easily identify key attributes correlated with a certain class. For example, in Fig. 3.15a, the feature "feathers" is attracted toward the class "bird." To facilitate the exploration process, the color of attributes is used to encode the most correlated label, and the radius encodes the magnitude of correlation with all the labels. For a more targeted exploration, when users hover over a specific class label, the radius dynamically encodes the correlation specifically with that class label. This interactive functionality improves the precision of attribute-class correlations and provides a more user-friendly and insightful experience for users.

In contrast to methods that visualize attributes directly, other VIS4AI methods visualize samples first and then assess the importance of each attribute in the visualization. For example, Wang et al. [252] proposed discriminative star coordinates to better identify important attributes in datasets. In traditional star coordinates, the k coordinate axes are uniformly arranged on a circle with the origin at the center, and a $2 \times k$ projection matrix $G = [g_1, g_2, \ldots, g_k]$ is used to project samples from the $k-$dimensional space onto a 2D space, i.e., $x_{\text{low-dim}} = G \cdot x_{\text{high-dim}}$. Here, each g_i serves as the anchor point of i-th dimension, and samples are projected as a linear combination of these anchor points. In discriminative star coordinates (Fig. 3.15b), the anchor points of dimensions are calculated by maximizing between-class scatter and minimizing within-class scatter. This strategy results in an enhanced separation between classes. Consequently, the distances of anchor points from the circle's origin reflect the importance of the corresponding attributes. If an attribute is important and highly correlated to classes, its anchor points will be positioned further from the origin to enhance class separation. Conversely, attributes of lower importance will have anchor points placed closer to the origin. This projection allows users to easily identify less important attributes and remove them.

PCPs are also widely used to visualize high-dimensional data due to their capability to visualize all samples with multiple attributes in a compact form [106, 112, 233]. Users can analyze the attributes and then select important ones in this visualization. For example, Johansson and Johansson [112] developed an interactive dimensionality reduction tool. This tool starts by allowing users to select quality metrics and configure the weighting parameters

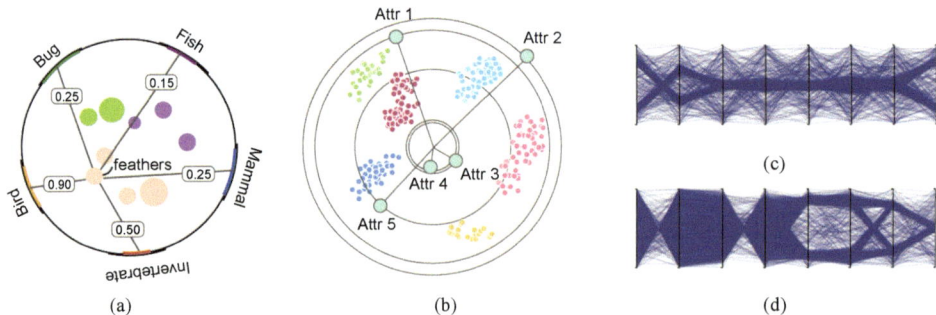

Fig. 3.15 Representative visualizations for analyzing attributes: **a** RadViz [9]; **b** discriminative star coordinates [252]; **c** interactive PCP [112] that enhances the cluster structures; **d** interactive PCP [112] that enhances the correlation structures

to combine multiple metrics. The attributes with high metrics are then kept and presented in PCPs for thorough analysis. To enhance the perception of patterns such as cluster structures or correlation structures, an ordering technique is employed to arrange the axes in PCPs based on user preferences (Fig. 3.15c and d). In addition, the quality metrics of each attribute are displayed below the corresponding axis, and the correlations between two adjacent attributes are positioned between the corresponding axes. This additional information facilitates the understanding of the quality and relationships of multiple attributes.

3.4 Summary

In this chapter, we classify existing VIS4AI studies for improving data quality into three categories: instance diagnosis, annotation diagnosis, and feature engineering. The first category facilitates the examination of the distribution and content of instances, thereby enabling them to make informed adjustments on individual instances. The second category supports correcting inaccurate annotations, supplementing insufficient annotations, and refining inexact annotations to enhance the accuracy and comprehensiveness of supervision for ML models. The final category focuses on interactively optimizing the input features of a model by adding and/or removing features, which ensures that the most relevant and informative input features are used in model training.

There are also many other VIS4AI studies to achieve the three tasks for better data quality, which cannot be introduced one by one in detail. Table 3.1 provides a concise summary of them based on the three data quality-related tasks.

Table 3.1 Summary of the VIS4AI studies in the data preparation stage

	Inaccurate	Insufficient	Inexact
Instance	[5, 7, 15, 71, 72, 83, 84, 116]	[38, 82, 92, 130]	[74, 207, 219, 253]
Annotation	[3, 13, 157, 190, 193, 194, 260, 267]	[37, 87, 89, 93, 96, 103, 110, 113, 120, 127, 137, 176, 210, 222, 223, 225, 265]	[39, 40]
Feature		[31, 35, 156]	[9, 67, 106, 112, 178, 214, 231, 233, 252, 277],

Techniques for Model Development

4

VIS4AI drives the development of ML models through three main tasks [44, 154, 270], model understanding, model diagnosis, and model steering. **Model understanding** aims to explain the rationale behind ML model predictions and their inner workings, and make these complex models at least partially understandable. **Model diagnosis** is the process of identifying and addressing defects or issues within ML models that fail to converge or do not achieve acceptable performance. **Model steering** is a method of interactively incorporating expert knowledge and expertise into the improvement and refinement process of an ML model, which is achieved by combining a set of rich user interactions with techniques such as semi-supervised learning or active learning. Various visualization techniques, such as node-link diagrams, scatterplots, and parallel coordinate plots, have been adopted to support these three tasks. Figure 4.1 shows the connections between ML models, the three VIS4AI tasks, and different visualization techniques.

4.1 Model Understanding

ML models often function as black boxes due to the lack of transparency regarding their internal working mechanisms. The increasing concerns about model trustworthiness and fairness make the black-box nature of ML models a significant issue for model developers and accordingly lead to many studies on model understanding. Focusing on the visualization techniques employed, this chapter exemplifies how VIS4AI facilitates the process of understanding powerful but complex ML models.

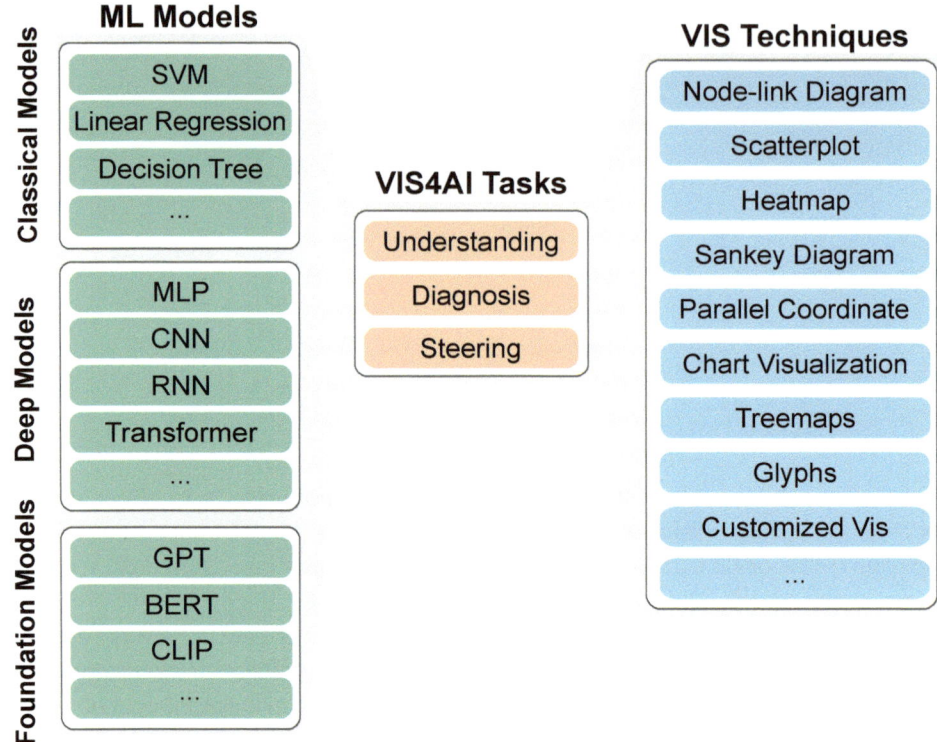

Fig. 4.1 An overview of the categories of ML models, VIS4AI tasks, and visualization techniques for model development

4.1.1   Node-Link Diagrams

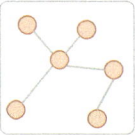

Node-link diagrams use 2D nodes to represent individual entities and links between the nodes to denote their connections. Additional information can be further encoded into different visual channels of the nodes and links such as node size, node color, and link style. This visualization has been extensively used to explore the structure of deep models and illustrate how different parts of the models, such as neurons or layers, cooperate to produce the final predictions.

Earlier efforts have employed directed graph layouts to visualize the structure of simple multi-layer perception (MLPs). For example, Tzeng and Ma [234] used a three-layer MLP to judge whether voxels from a 3D volume MRI dataset represent brain materials. The structure of the MLP is visualized as a node-link diagram, where the nodes and links represent the neurons and the connections between neurons, respectively. Regarding the layout, the MLP layers are arranged horizontally with uniform spacing, and the neurons within each layer are represented as nodes and arranged vertically with equal distance between them. Neurons from neighboring layers are further connected with links to denote the connectivity between layers. The color of the nodes and the width of the links encode the values of the corresponding neurons and weights, respectively. The visualization effectively discloses the important input features contributing to the output predictions and helps to understand the MLP model. However, it also suffers from severe scalability issues. The visualization gets cluttered easily when the model structure grows in complexity, such as in the case of a deep neural network (DNN) with hundreds of layers/neurons or tens of thousands of neuron connections.

To address this issue, Liu et al. [149] developed CNNVis (Fig. 4.2) to visualize deep convolutional neural networks (CNNs). This visual analytics tool leverages clustering tech-

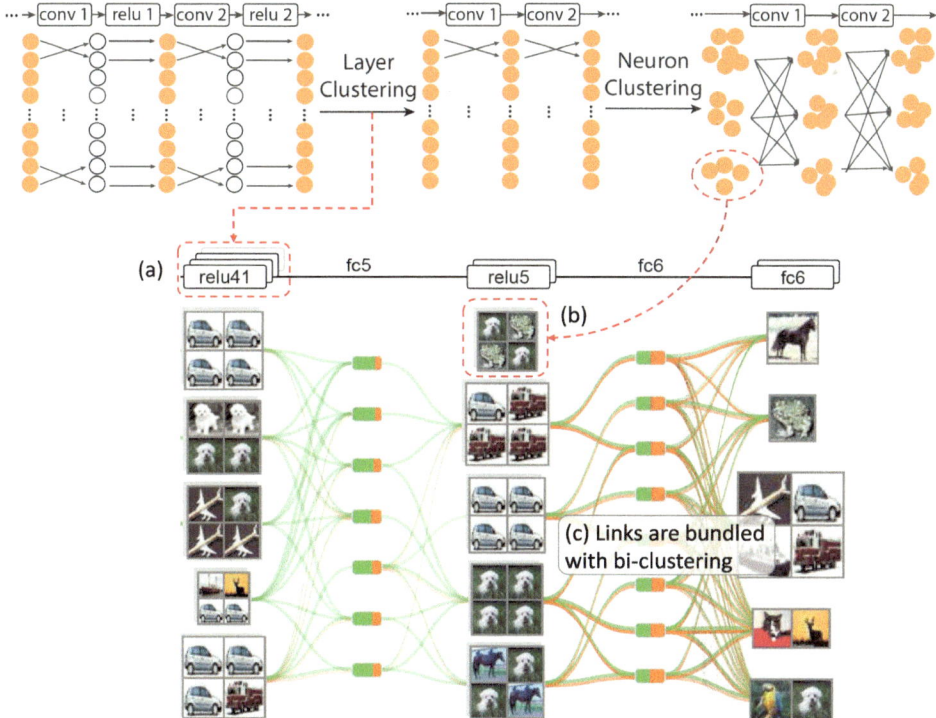

Fig. 4.2 The node-link diagram designed in CNNVis [149]: **a** merging neural layers between consecutive pooling layers of a CNN; **b** grouping neurons with similar roles and presenting them with image patches that activate the neurons the most; **c** bundling connections between neighboring layers with bi-clustering

niques at three levels to reduce visual clutter. The first level greatly reduces the number of layers needed to visualize by merging neural layers between consecutive pooling layers (Fig. 4.2a). Merging these similar layers does not hinder the understanding of a CNN, as (1) the outputs from these layers have one-to-one correspondences, and (2) domain experts often care more about the final activations instead of the intermediate convolutional results. The second level aggregates similar neurons in the same layer into clusters and visualizes each cluster using image patches that maximally activate the corresponding neurons (Fig. 4.2b). As the number of neurons/patches in each cluster could be large, but the space to visualize them is limited, a hierarchical rectangle packing algorithm is introduced to organize the patches hierarchically and place them compactly. The third level utilizes bi-clustering to reduce the visual clutter of links between layers. As shown in Fig. 4.2c, the resulting bi-clusters are represented as "in-between" nodes [230] with green and orange rectangles whose width represents the numbers of the edges with positive and negative weights, respectively. With multiple coordinated visualizations, CNNVis helps users understand the roles of neurons and their learned features. Moreover, it illustrates how low-level features are aggregated into high-level ones through the network.

Many other prominent studies in VIS4AI utilize node-link diagrams to visualize the architecture of DNNs. For example, Wongsuphasawat et al. [256] designed a graph visualization to explore the architecture of ML models in TensorFlow through a series of graph transformations. Kahng et al. [114] employed a node-link diagram to summarize the DNN's computation graph in ActiVis. TensorFlow Playground [221] visualizes a DNN directly as a node-link diagram. This visualization allows users to interact with the layers and neurons of the DNN. Additionally, node-link diagrams have also been widely used to visualize the structure of tree-based models. In these visualizations, nodes represent internal or leaf tree nodes, and links denote the parent-child relationship between nodes [156, 243, 264]. These visualizations facilitate an intuitive understanding of the hierarchical structure of tree models, as well as the data flow through different tree branches.

4.1.2   Scatterplots

Scatterplots present data samples as 2D points and encode information within the 2D coordinates of those points. Different visual channels of points, such as their color, size, and shape, can be used to encode user-interested data features. Scatterplots are frequently used to reveal similarity or dissimilarity relationships between data samples and their clustering patterns.

Li et al. [141] utilized the power of scatterplots to illustrate the attention patterns of different attention heads in vision transformers (ViTs). A ViT typically consists of many attention heads organized by a hierarchy of attention layers. Different heads encode different attention patterns between image patches and contribute differently to model performance. Revealing the attention patterns inside individual heads and comparing them across heads from different layers are crucial for understanding ViTs. Assuming that the total number of attention heads in a ViT is m, feeding n images into the ViT will generate $m \times n$ heads. Each comes with an attention matrix sized $p \times p$, where p denotes the number of patches into which each image has been decomposed, and the matrix reflects the pairwise attention between all patches. These $m \times n$ attention matrices are projected as a 2D scatterplot using t-SNE. As shown in Fig. 4.3a, the $m \times n$ heads can be classified into two categories: (1) smaller isolated clusters and (2) a larger central cluster.

Figure 4.3b illustrates the attention head of an image (head 11, layer 1) in the first category. In this head, each patch consistently attends to the patch on its right. This creates a pattern in the attention matrix that shifts one cell off the diagonal. The "gaps" in the attention matrix indicate that the rightmost patches of the original image lack a subsequent patch to attend to. This head exhibits a consistent attention pattern across all n images, regardless of their content. As a result, it is *content-agnostic*, and its attention matrices from all n images form an isolated cluster in the scatterplot. Figure 4.3c, d, and e show another three content-agnostic

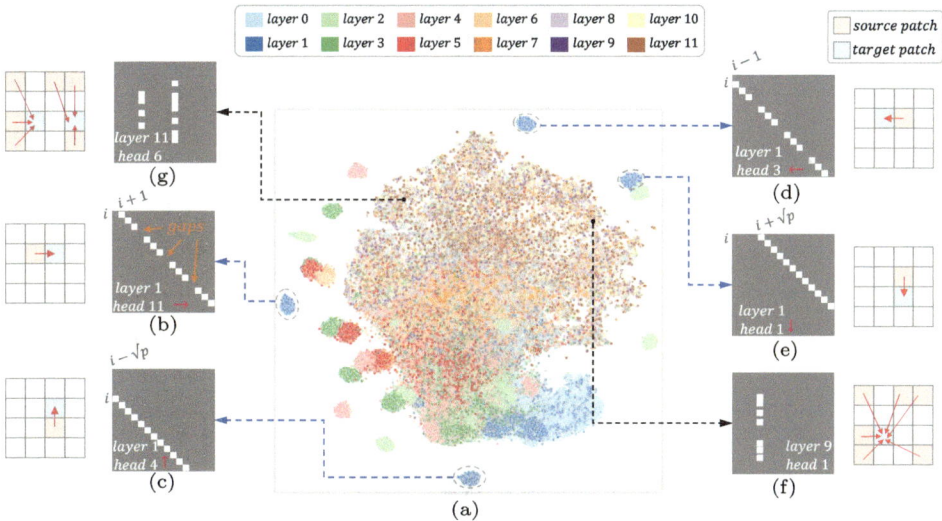

Fig. 4.3 Understanding the attention patterns learned by individual heads of a ViT: **a** each point in the scatterplot represents an attention head, and its color denotes the attention layer; **b, c, d, e** these four heads demonstrate content-agnostic attention patterns that make each patch attend to the patch on their right, top, left, and bottom respectively; **f, g** these two heads present content-relevant attention patterns

heads, in which each image patch attends to the patch on its top, left, and bottom, respectively. Figure 4.3f shows an attention head in the second category. The vertical line pattern in the attention matrix indicates that many patches (matrix rows) direct their attention toward one important patch (matrix column). For example, when processing an image of a cat, this head may direct all background patches to focus on the patch containing the cat. When processing two cat images with the cat in different positions, the vertical line pattern shifts to different columns in the attention matrices. This indicates that the pattern in the attention matrix of this head is *content-relevant*. Figure 4.3g illustrates another content-relevant head, where two important target patches and two vertical line patterns are observed in the attention matrix. Since the important patch may appear in different parts of the image, the points representing the attention matrices of different images scatter throughout the space. The spread of these points results in the large central cluster in Fig. 4.3a.

In the scatterplot, the color of each point denotes the attention layer from which the represented head comes. The color distribution of the points reveals that *content-agnostic* heads are often from lower layers (layer 0 to layer 5), whereas *content-relevant* heads are more common in higher layers (layer 6 to layer 11). This pattern reveals different learning heuristics developed by the heads in different layers. The lower layer heads use fixed attention patterns to extract basic features from the images, whereas the higher layer heads combine important basic features to learn object-level information.

Beyond a single scatterplot, multiple scatterplots can be presented side-by-side to show the dynamic evolution of a DNN or compare the data representations from different ML models. For example, Rauber et al. [203] employed a set of scatterplots to depict the evolution of a DNN's hidden states during its training. They used the MNIST test set and an MLP to illustrate the idea. Figure 4.4a shows the t-SNE projection of the last hidden-layer activations for 2000 test images before training the MLP. Each point represents an image and is colored by its class label. Figure 4.4b shows the projection of these activations after training the MLP for 100 epochs. By comparing these two scatterplots, especially the cluster patterns of the images of different classes, the authors disclosed how the MLP gradually gains the power to separate instances of different classes. Furthermore, by examining the misclassified images, the authors also found some quality issues associated with the MNIST images. For example, the image at the top right of Fig. 4.4b is an image of digit 3. However, due to its similarity to a digit 5, it is misclassified as digit 5 and falls into the gray cluster of digit 5. Similar visual outliers can also be highlighted by encoding the corresponding points with different shapes, such as the outliers represented with triangles in Fig. 4.4c.

There are more VIS4AI studies using scatterplots to visualize the dimensionality reduction results of high-dimensional data from ML models. For example, Wang et al. [241] employed a scatterplot to visualize the latent image representations of a CNN classifier. The overview provided by the scatterplot helps select image samples around the decision boundary for instance-level model diagnosis. Similarly, Kahng et al. [114] used a scatterplot to visualize the DNN activations of textual data. Li et al. [139] developed EmbeddingVis, a visual analytics tool that employs multiple scatterplots to display graph embeddings of the

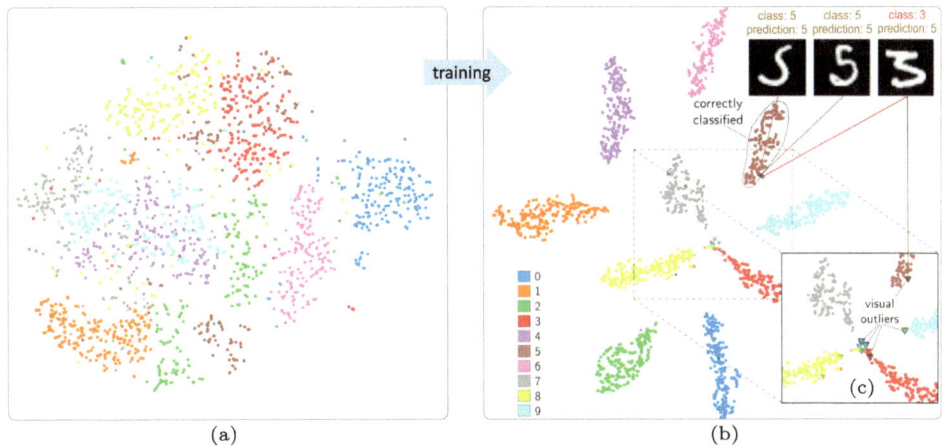

Fig. 4.4 Using dimensionality reduction algorithms and scatterplots to visualize the high-dimensional activations from DNNs: **a** the last layer activations of an MLP before any training; **b** the same set of activations after the model has been trained for 100 epochs; **c** encoding the points representing outliers with triangles

same set of nodes, each generated by a different ML model. To compare these embeddings and relate the views, the same nodes across scatterplots are connected with explicit edges.

4.1.3 Parallel Coordinate Plots (PCPs)

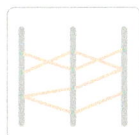

PCPs use multiple parallel axes to represent different dimensions of high-dimensional data, one for each dimension. A high-dimensional data instance is represented by a polyline (or a curve) connecting the values across all parallel axes. Superimposing multiple instances on these axes reveals their collective behaviors across dimensions and thus highlights patterns, correlations, and outliers.

LSTMVis [227] employs a PCP to visualize a large number of high-dimensional hidden states generated by long short-term memory (LSTM) models. These models take a sequence of t tokens as input, where each token is encoded as a k-dimensional hidden state. This encoding utilizes information from the token itself along with the hidden state of its preceding token. A key area of investigation often focuses on understanding the specific information encoded within individual hidden-state dimensions. LSTMVis represents each token as a parallel axis and marks its k-dimensional hidden state as k values on the axis. Connecting

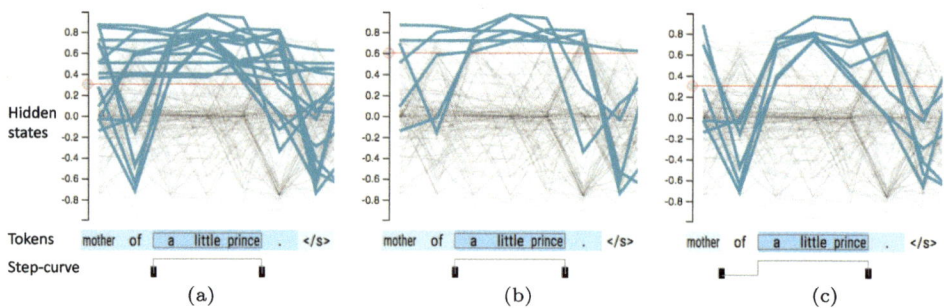

Fig. 4.5 LSTMVis [227] employs a PCP to interpret the hidden states of LSTM models: **a** filtering hidden-state dimensions whose values for tokens "a little prince" are above the threshold 0.3, indicated by the red horizontal line; **b** increasing the threshold to 0.6 results in fewer polylines; **c** extending the pattern by specifying the hidden states for the token "of" to be off

the values of the same hidden-state dimension across tokens forms a polyline. In total, there are k polylines. Simply superimposing them in the plot may result in visual clutter, which prevents users from discovering the functionality of each hidden-state dimension. To address this problem, interactive pattern filtering is introduced in LSTMVis. As shown in Fig. 4.5a, the blue-shaded polylines are the hidden-state dimensions with values for the tokens "a little prince" above the user-specified threshold 0.3, marked by the red horizontal line. The remaining hidden-state dimensions fade off with thinner polylines and a gray color. As shown in Fig. 4.5b, increasing the threshold to 0.6 narrows the focus to fewer polylines. Furthermore, users have the flexibility to filter the hidden-state dimensions of the tokens of interest, which are specified by the step curve at the bottom of the view. For example, compared to Fig. 4.5c applies additional filtering to exclude hidden-state dimensions whose values for the token "of" fall below the threshold.

SCANViz [242] explains the symbol concept association network (SCAN) model by revealing how it uses variational autoencoders (VAEs) to encode and decode visual concepts and associates visual concepts with different text symbols. The work utilizes a PCP to depict the 32D latent space of a VAE and explain what concepts individual latent dimensions have encoded. In this visualization, each axis represents a latent dimension (Fig. 4.6b). One polyline connecting values over the 32 axes represents the latent representation of one image. Feeding the 16 images of a 3D scene with varying floor colors in Fig. 4.6a into the VAE encoder and visualizing the variance of their latent representations on each latent dimension (the red region in the PCP), one can clearly see that dimensions "d18," "d20," and "d21" have obviously larger variances than others. It is hypothesized that these three latent dimensions mainly control the floor color of the images. To verify this, in Fig. 4.6c, users perturb the latent value on "d20" and then reconstruct the perturbed representations using the VAE. The main difference in the six reconstructed images comes from their floor color. This verifies

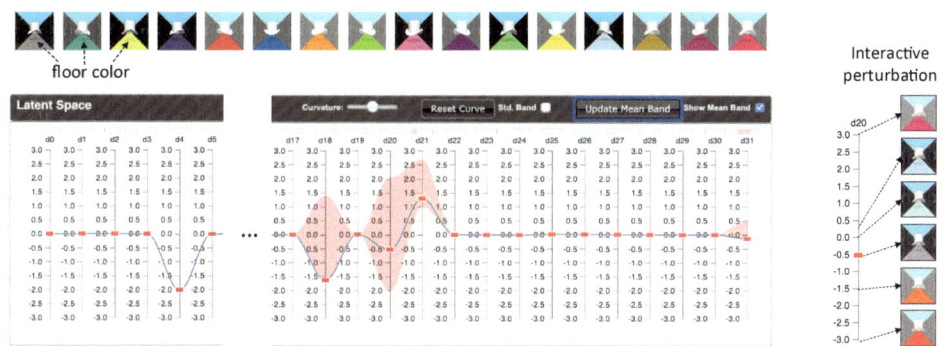

Fig. 4.6 SCANViz [242] uses a PCP to explore the latent dimensions of a VAE: **a** 16 images of a 3D scene with varying floor colors; **b** feeding the row of images into the VAE encoder and observing the variance (the red area) of individual latent dimensions through a PCP; **c** perturbing the value on dimension "d20" and verifying its controlled visual concept through the reconstructed images

the previous hypothesis that "d20" controls this visual concept. A similar perturbation can also be performed on dimensions "d18" and "d21" to verify that they control the floor color.

In addition to LSTMVis and SCANViz, several other VIS4AI methods utilize PCPs to visualize the latent space of an autoencoder (AE), VAE, or their variations. For example, DeepVID [241] employs a PCP to visualize the latent space of a VAE and conducts data perturbation in the space for data augmentation. VATLD [82] and VASS [92] apply PCPs to visualize the latent space of a β-VAE and a conditional-VAE, respectively. The visualizations help interpret and disentangle the encoded visual concepts in the models' latent space.

4.1.4   Heat Maps

Heat maps represent complex information in a 2D, colored matrix format. Each cell within the matrix is colored to represent the magnitude or intensity of a particular variable, making it easier to see patterns, trends, and outliers. This type of visualization can effectively present high-dimensional data from different stages of ML models and reveal hidden data patterns.

Karpathy et al. [117] studied character-level language modeling with a recurrent neural network (RNN), and employed heat maps to investigate what syntax and semantic information the RNN's hidden states have captured. The RNN processes a sequence of characters with the specific goal of predicting the subsequent character based on the preceding ones. The intermediate hidden states of the RNN can be considered as a $t \times k$ sized matrix, where

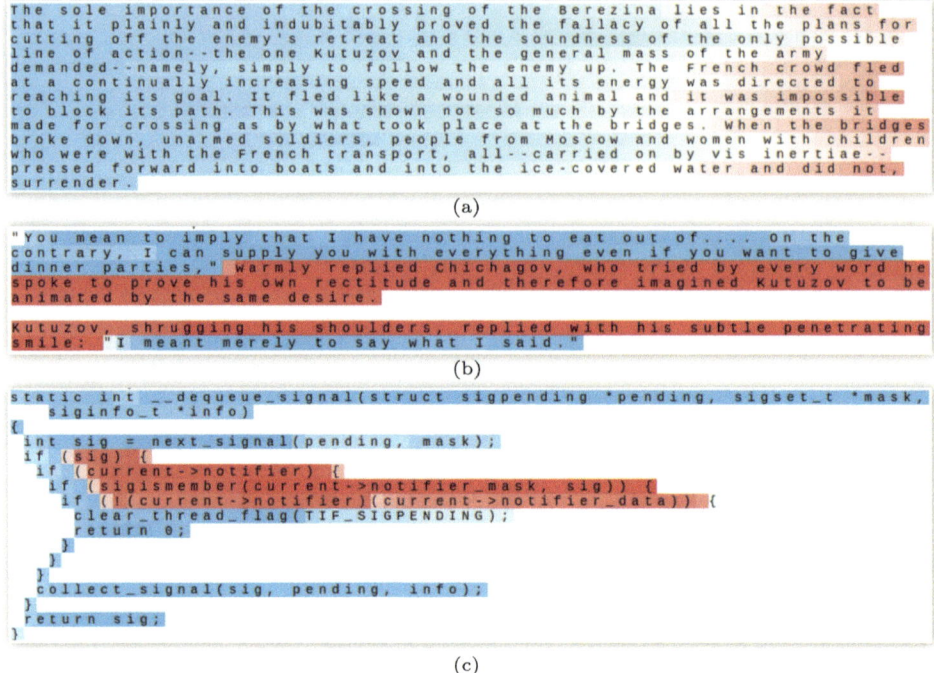

Fig. 4.7 Interpreting RNN's hidden-state dimensions with heat maps [117]: **a** the dimension that captures the position of characters in individual text lines; **b** the dimension that signifies characters inside quotations; **c** the dimension that captures the characters inside the if-statements of the code

t is the number of characters in the input sequence, and k is the dimensionality of the hidden states. Each dimension of the hidden states typically encodes different information from the input. According to the above formulation, an RNN hidden-state dimension can be considered as a t-length vector, in which each element represents the dimension's response level to the corresponding input character. Visualizing individual hidden-state dimensions as heat maps and associating them with the corresponding input characters can effectively reveal what semantics the dimensions have captured. Figure 4.7 visualizes three hidden-state dimensions of an RNN. For each dimension, several input sequences are presented, and the background of individual characters is colored using values from the corresponding hidden-state dimension. The values of RNN hidden states range from -1 to 1 and are encoded by a diverging-sequential color scheme from red to blue. As shown in Fig. 4.7a, the hidden-state dimension works as a position indicator of characters, and its value decreases from the beginning to the end of each text line. The hidden-state dimension in Fig. 4.7b gets activated only for characters within quotations. The hidden-state dimension in Fig. 4.7c extracts the conditions of "if-statements" from the code pieces. These heat maps relate the input

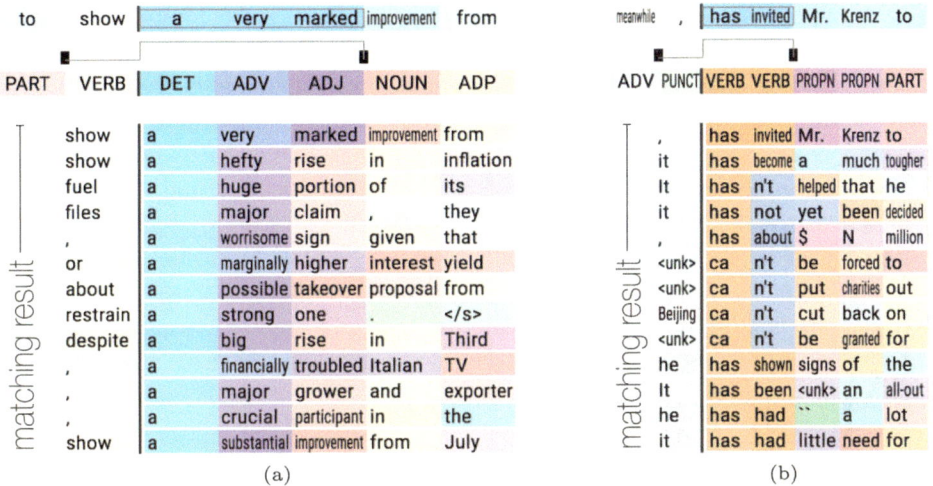

Fig. 4.8 LSTMVis [227] employs heat maps to explore the relationships between LSTM hidden states and the POS of input words: **a** input phrases with similar hidden-state patterns to the specified noun phrase are mostly noun phrases; **b** instances with similar hidden-state patterns to the specified verb phrase are mostly verb phrases

sequences to the internal hidden states, disclosing what individual hidden-state dimensions have learned.

LSTMVis [227] is another typical example of heat map visualization. To convincingly reveal what LSTMs' hidden states have encoded, LSTMVis provides an interface for users to define an active pattern of the hidden states. The tool then identifies input instances with similar hidden-state patterns and presents the matching results with a heat map to augment their common pattern. Figure 4.8a shows an example of analyzing an LSTM trained for next-word predictions. The top of the view defines a sequential pattern to be examined, which has the part-of-speech (POS) sequence DET-ADV-ADJ-NOUN. The bottom of Fig. 4.8a shows sentences with similar hidden-state patterns as a heat map. Each row is an input sentence, and sentences are aligned vertically based on their POS pattern. The color of each heat map cell encodes the POS of the corresponding word (DET: cyan, ADV: blue, ADJ: violet). As the colors in individual heat map columns are mostly similar, the matching results augment the pattern of the selected noun phrase. This helps users confirm their hypothesis that the LSTM learns the syntax of language structures. Figure 4.8b shows another example in which the specified pattern is a verb phrase. From the heat map, the instances with matched patterns are also verb phrases. These heat maps help interpret the model by confirming or rejecting users' hypotheses on what the ML models have learned.

There are more VIS4AI studies that employ heat maps to interpret ML models [85, 91, 107, 172]. For example, RNNVis [172] clusters a large number of high-dimensional RNN hidden states and visualizes each cluster as a "memory chip" through a heat map, where the color of the heat map cells denotes the response level of the corresponding hidden states. Users can gain insights into the behavior of the RNN hidden states from the color patterns of these memory chips in response to different input words. Furthermore, heat maps are also widely used to interpret RNN-based deep reinforcement learning (DRL) models. For example, both DRLViz [107] and DynamicsExplorer [91] use heat maps to externalize the memory space of DRL agents. The large number of input tokens and the high dimensionality of the hidden states may cause the heat map to occupy significantly more screen space. Jaunet et al. [107] addressed this problem by enabling interactions to zoom in/out of a large heat map. He et al. [91] applied principle component analysis (PCA) to reduce the dimensionality of hidden states first and then used heat maps to visualize the top-k principle components only. These improvements make heat maps easily scalable to large or high-dimensional data.

4.1.5 Glyphs

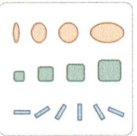

Glyphs are symbolic/iconic representations of complex data, which encode one or multiple data attributes through various visual channels of a geometric object such as its shape, size, and angle [211]. These glyphs have been used to represent different types of ML data, including different features of the model input, varying aspects of the model output, and different model architectures.

DynamicsExplorer [91] designs customized glyphs to illustrate the input states and output actions of a DRL model. The model is trained to play a game called "ball-in-maze" (Fig. 4.9a), where the goal is to guide the ball from the outer rings to the center of the maze by tilting the maze. Figure 4.9b shows the simulated game scene. The five possible agent actions include tilting the maze clockwise or counter-clockwise around the x-axis (x+, x-), y-axis (y+, y-), or no action (Fig. 4.9c). Figure 4.9d presents a sequence of glyphs specifically designed to represent game states and agent actions. Each glyph in the sequence uses a point, a dashed circle, and an arrow to encode the position of the ball, the wall of the inner ring, and the accumulated tilts from all directions, respectively. The color of the point and the arrow represent the action issued by the DRL model, as indicated by the legends in Fig. 4.9c. These glyphs intuitively convey the dynamic tilting of the maze in response to the movement of the ball, offering insights not apparent from the static trajectory shown in Fig. 4.9b.

GNNLens [111] interprets the importance of node features and edge connections in a GNN classifier using two proxy models. The first model, an MLP, considers only the graph node features while disregarding node connectivity. Conversely, the second model, a GNN,

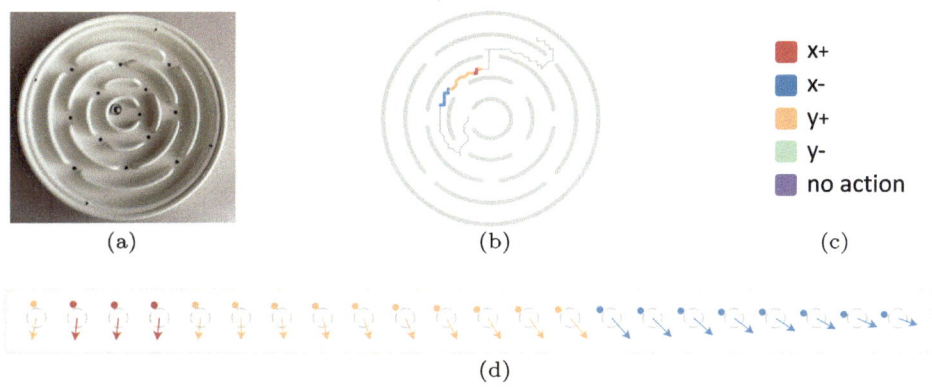

Fig. 4.9 Interpreting a reinforcement learning model with DynamicsExplorer [91]: **a** the "ball-in-maze" game; **b** the simulated game environment; **c** the five possible actions for the reinforcement learning agent; **d** a sequence of the designed glyphs that encodes input game states and output agent actions

focuses solely on node connectivity while disabling the node features. To simultaneously present the prediction results from the original GNN and the two proxies, a circular glyph is designed for each graph node. As shown in Fig. 4.10a, the colors of the inner circle and the three outer arcs denote the true label and the predicted labels from the three models, respectively. This glyph makes it easier to understand how a graph attention network classifies paper types in a citation graph. In this graph, the nodes represent individual papers, each belonging to one of the seven possible types shown in Fig. 4.10b. The links denote the citation relationships between the papers. A node-link diagram (Fig. 4.10c) visualizes the graph for an overview, where the focused graph node is encoded by the designed glyph to reflect predictions from the three models. By tracing the focused node in the node-link diagram, one can examine the predictions for its 1-hop and 2-hop neighbors through the corresponding glyphs.

Glyphs have also been widely used to explain the architecture components of different ML models [14, 245, 271]. For example, Wang et al. [245] developed DNN Genealogy as an educational tool to summarize the possible DNN architectures. As the DNN architectures from different research publications are often presented in different styles, it is challenging to compare them. To address this issue, the authors designed a set of network glyphs with a consistent style and utilized them to visualize the architectures of different DNNs. Following the principles that a network glyph should be concise, intuitive, and general, four basic network components are designed for (1) layer/block, (2) combination, (3) gate, and (4) connection (Fig. 4.11a). These glyphs are then assembled to generate the architecture diagram of any DNNs. Figure 4.11b shows an example architecture for DNNs with linear gated units. These glyphs effectively depict key features of various DNNs, which simplifies the comprehension and comparison of continuously evolving DNN architectures.

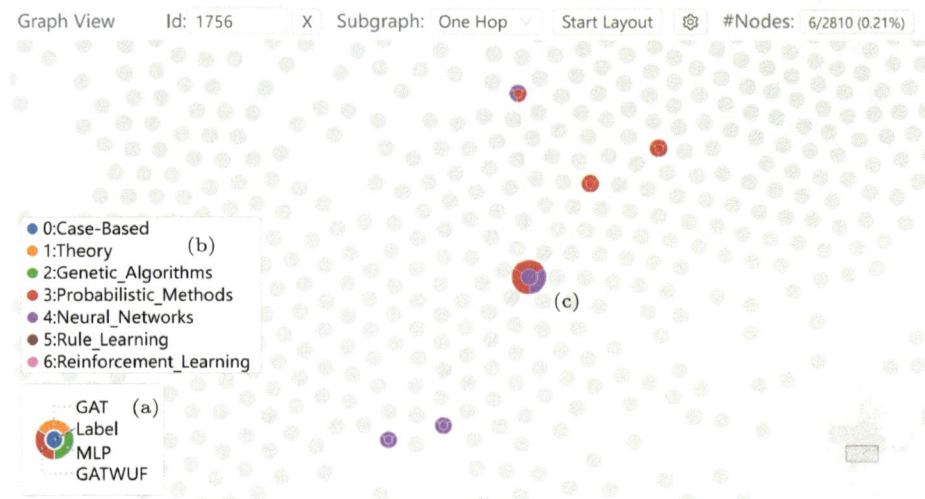

Fig. 4.10 Interpreting a graph attention network with GNNLens [111]: **a** the designed glyph for each graph node; **b** the seven categories of the graph nodes; **c** the graph is visualized through a node-link diagram

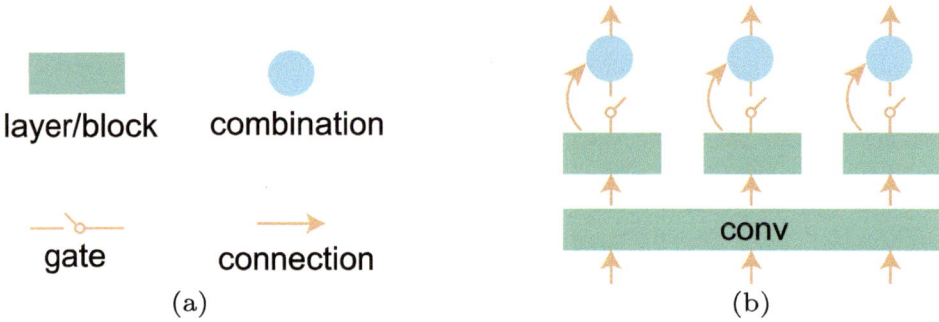

Fig. 4.11 Network glyphs: **a** the four key network components; **b** linear gated units built upon the four key components

In summary, an increasing number of VIS4AI studies are utilizing glyphs to represent various aspects of ML models [16, 53, 172]. In addition, researchers often integrate these glyphs with other visualization techniques, such as node-link diagrams [163, 164] and scatterplot visualizations [165, 250], to enhance the visualization of diverse information. This integration not only enriches the visual presentation but also promotes a clearer understanding of complex data relationships.

4.2   Model Diagnosis

In many cases, complex ML models do not perform as expected and produce incorrect results. VIS4AI addresses this issue by creating interactive visual analytics tools. These tools enable users to analyze the log data of ML models, facilitating the identification and diagnosis of malfunctions. This chapter exemplifies typical model diagnosis studies and illustrates the roles that different visualizations have played.

4.2.1   Chart Visualizations

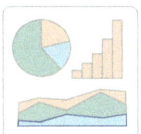

Chart visualizations, including line charts, bar charts, and pie charts, compile diverse information types for user-friendly presentation. Different charts have their respective merits in demonstrating information in different formats. Combining these charts harnesses their collective capabilities to effectively depict the large and multi-faceted data from ML models, which empowers model developers in diagnosing abnormal model behaviors.

DQNViz [240] employs stacked area charts, pie charts, stacked bar charts, and line charts to monitor the training statistics of a deep Q-learning network (DQN) agent trained for the Atari Breakout game (Fig. 4.12a). In this game, the agent controls the paddle at the bottom of the scene through four actions: "moving left," "moving right," "no operation," and "firing the ball." The goal is to catch the ball and destroy the six layers of bricks at the top of the scene. Each brick from the bottom, middle, and top two layers corresponds to a 1-point, 4-point, and 7-point reward, respectively. Through a stacked area chart extended horizontally from left to right, DQNViz tracks the count of rewards at each training epoch. Figure 4.12b shows a total of 200 training epochs, where the light green, purple, and orange colors represent 1-point, 4-point, and 7-point rewards, respectively. Initially, the 1-point rewards dominate the stacked area chart. This indicates that the agent is not adequately trained and tends to destroy bricks from the lower layers. As training progresses, the frequency of 4-point and 7-point rewards gradually increases and matches the count of 1-point rewards around epoch 120. At this stage, the agent has been significantly improved and is capable of destroying bricks from the six layers. After epoch 120, the counts of the three types of rewards remain similar, reflecting the good performance of the DQN agent. Overall, the stacked area chart depicts a successful training of the model.

However, a detailed examination of the stacked area chart reveals certain epochs with abnormal reward distributions such as epoch 37 in Fig. 4.12b. The light green area dominates the distribution in this epoch. Selecting the epoch displays its reward distribution using a pie chart (Fig. 4.12c1). Meanwhile, the reward distribution inside individual episodes of this

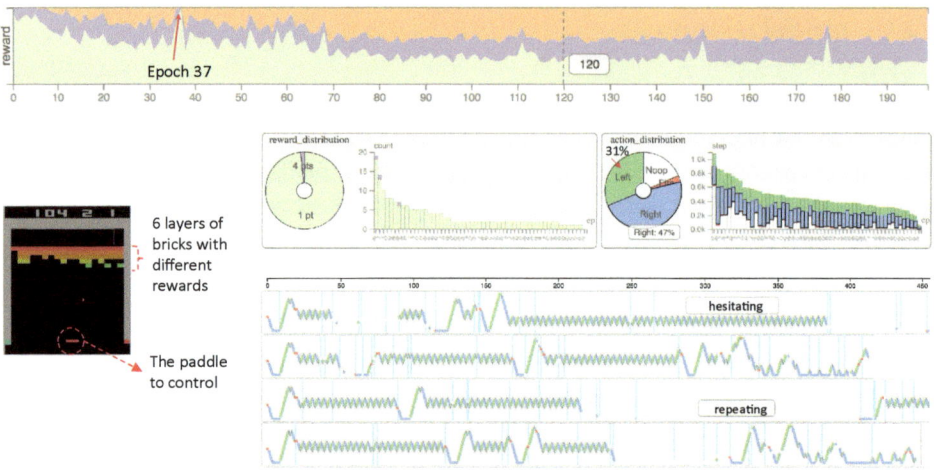

Fig. 4.12 Diagnosing DQN with DQNViz [240]: **a** the Atari Breakout game; **b** monitoring the DQN training process with a stacked area chart; **c** the reward distribution in epoch 37; **d** the action distribution in epoch 37; **e** the Trajectory view reveals two abnormal agent moving patterns

epoch[1] is presented through a stacked bar chart (Fig. 4.12c2), one bar for one episode. It can be seen from these two charts that the agent in this epoch mostly gets 1-point rewards. This indicates that it mainly destroys bricks from the bottom two layers. In Fig. 4.12d, another set of a pie chart and a stacked bar chart shows the action distribution in this epoch. These two charts show that the agent mainly takes actions "moving left" (31%) and "moving right" (47%). The length of individual episodes is not short, but the accumulated reward is very low. This indicates that the agent frequently executes actions that do not lead to significant rewards. The agent behavior can be further examined in the trajectory view (Fig. 4.12e). This view presents each game episode as a line chart, where the x-axis denotes the game step, and the y-axis denotes the paddle's distance to the right boundary of the scene. The color of the line encodes the actions issued by the agent in different steps. Four episodes within the selected epoch are presented and arranged vertically. From them, two dominant agent behavior patterns emerge. The first is the "hesitating" pattern, in which the agent frequently switches between "moving left" (green) and "moving right" (blue), without the ball in the scene (Fig. 4.12e1). This indicates that the agent fails to recognize the need to fire the ball first before it can bounce to destroy bricks. This behavior also explains the large portion of "moving left" and "moving right" actions in the pie chart of Fig. 4.12d. The second is the "repeating" pattern, in which the agent continues repeating the "no operation" action (white) to stay in the scene, but does nothing (Fig. 4.12e2). Although these two patterns help the agent survive longer in the game, they do not contribute to learning brick-destroying

---

[1] Each epoch contains a fixed number of agent steps, which covers many game episodes. Each game episode consists of the steps from the start of a game to its end.

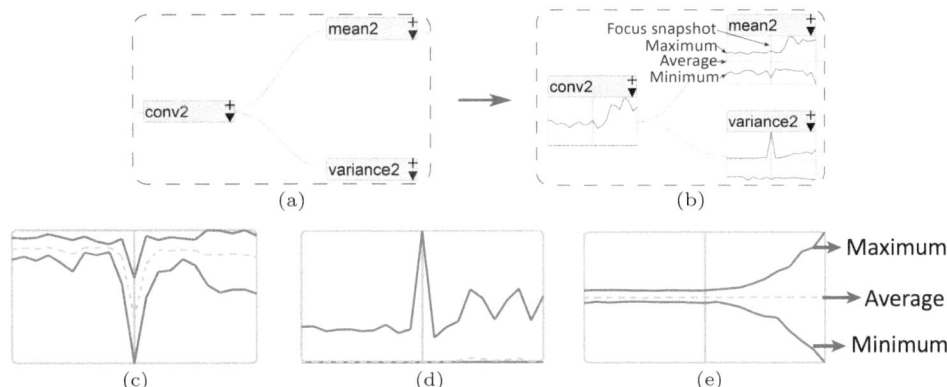

Fig. 4.13 Diagnosing DGMs with DGMTracker [151]: **a** the architecture of a DGM is formulated as a DAG and visualized using a node-link diagram; **b** each node of the diagram is a neural layer and the aggregated statistics of the layer within a time period are visualized as a line chart; **c, d, e** different patterns in the line charts disclose abnormal model behaviors

strategies. Disclosing these patterns to users assists in diagnosing abnormal agent behaviors. In this particular case, users can adjust different hyperparameters, e.g., random action rate, based on the revealed details to reduce the undesired agent behaviors and boost learning performance.

Another example that uses chart visualizations for model diagnosis is DGMTracker [151]. This tool formulates the network architecture of a deep generative model (DGM) as a directed acyclic graph (DAG). Figure 4.13a visualizes the DAG as a node-link diagram where the nodes represent layers and the links are the connections between layers. Each node utilizes a line chart to illustrate the data flow of the associated neural layer over a specific time period (Fig. 4.13b). Specifically, the central vertical line inside each node marks the focus snapshot S_t. The curves depict the training dynamics around the focus snapshot, including the maximum, average, and minimum activations in the snapshots from S_{t-k} to S_{t+k}. The line charts from different nodes disclose the training dynamics of the entire DGM, which helps diagnose abnormal model behaviors. For example, the line chart in Fig. 4.13c indicates that the focus snapshot experiences an abrupt decrease in activation values. Figure 4.13d shows the case that some activations increase abnormally in the focus snapshot, despite that most of the activations remain stable as the average curve is flat. These two cases suggest potential errors in the corresponding layers, which could lead to the pattern observed in Fig. 4.13e. In this pattern, the activations remain stable before the focus snapshot but become unstable after it. Besides activations, similar statistics for the gradients and weights of individual network layers can also be computed and plugged into the line charts to track the training dynamics of the model. These dynamics help monitor the model and identify the snapshots with abnormal behaviors for diagnosis.

Similar chart visualizations have been adopted in many other VIS4AI studies to debug or diagnose the corresponding ML models. For example, DriftVis [263] employs a line chart to track the drift degree of the input data and assess their quality for model training. Squares [204] utilizes variations of bar charts and line charts to show instance-level confusion in multi-class classification models. RNNbow [29] takes advantage of stacked bar charts to reveal issues, such as the gradient vanishing problem, in the training of RNNs.

4.2.2   Matrix Visualizations

Matrix visualization displays data or relationships in a structured matrix format. In a matrix, rows and columns intersect at cells. Each cell typically contains numerical data. The numerical values can be represented through different visual cues, such as color or shading, making it easier to detect patterns, correlations, or discrepancies within the data. In VIS4AI, matrix visualizations are commonly used for model performance analysis and causal reasoning.

Confusion matrices are often used to illustrate the class confusion of classification models. These matrices use rows to denote true classes and columns for predicted classes. Each cell counts the instances falling into the specific ground-truth class and predicted class combination. The numbers from all cells are then mapped to a visual channel such as color or shape. Since the ground-truth classes and predicted classes are often arranged in the same order along the rows and columns, the diagonal cells indicate correct predictions, while the non-diagonal cells show incorrect ones. This visualization, widely adopted in many visual analytics tools, shows the prediction power of classification models and facilitates the diagnosis of incorrect predictions [19, 42, 78, 145, 241, 258].

Bilal et al. [19] developed Blocks (Fig. 4.14) to dissect the hierarchical class confusion pattern within a confusion matrix. In large datasets, class labels are often organized by a hierarchical structure. For example, the "animal" class consists of both "vertebrate" and "invertebrate" animals. Further, the "vertebrate" sub-class can be divided into "mammal" and "bird", and the "mammal" sub-class includes sub-classes such as "dog", "cat", and "wolf". Reflecting class confusion at different levels of the hierarchy helps better understand the performance of a classification model and diagnose its confusion patterns. In Blocks, a horizontal icicle plot is employed to show the class hierarchy (Fig. 4.14a). Following the class order in the hierarchy, the confusion matrix (Fig. 4.14b) depicts class confusion at different levels. The value of 1 and the maximum value are respectively assigned to a light and a dark shade. Values in between are mapped to colors using a sequential color scale. The cells with value 0 are mapped to white, which facilitates the identification of non-zero cells. The diagonal cells represent true positives, which usually have larger values but are

Fig. 4.14 Blocks [19] extends the confusion matrix to analyze class confusion patterns with hierarchical class labels: **a** a horizontal icicle plot shows the class hierarchy; **b** the confusion matrix with subgroups of classes presented as blocks in red squares; **c** image samples of the selected group of classes

also less important for misclassification analysis. Therefore, they are excluded from the visualization. The block patterns, marked by the red squares, reveal the intra-group and inter-group class confusion. This feature enables users to focus on a specific class group for detailed examination. Specifically, the shaded region in Fig. 4.14b indicates user interaction with all classes in the "bird" group. The right panel (Fig. 4.14c) of Blocks shows image samples whose labels belong to this group.

More recently, Chen et al. [41] developed Uni-Evaluator, which provides a unified model diagnosis method for multiple computer vision tasks, including classification, object detection, and instance segmentation. First, to address the scalability issues when the number of classes is large, Uni-Evaluator organizes the classes hierarchically and presents them as an indented tree (Fig. 4.15a). The summary statistics (e.g., precision, recall) of each class or cluster are displayed on the right side of the matrix. Second, it supports three modes for analyzing both classification and localization performance in a unified manner. The only difference between the three modes is the content within the matrix cells. The confusion mode is similar to the traditional confusion matrix, where the color represents the number of samples in classification tasks and the number of objects in detection or segmentation tasks (Fig. 4.15a). The size mode utilizes a pie chart with three sectors to summarize the sizes of predicted objects in each cell (Fig. 4.15b). The gray sector represents predicted objects with precise sizes, while the yellow/green sector represents those with larger/smaller sizes compared to ground-truth (Fig. 4.15c). The angle of each sector is proportional to the num-

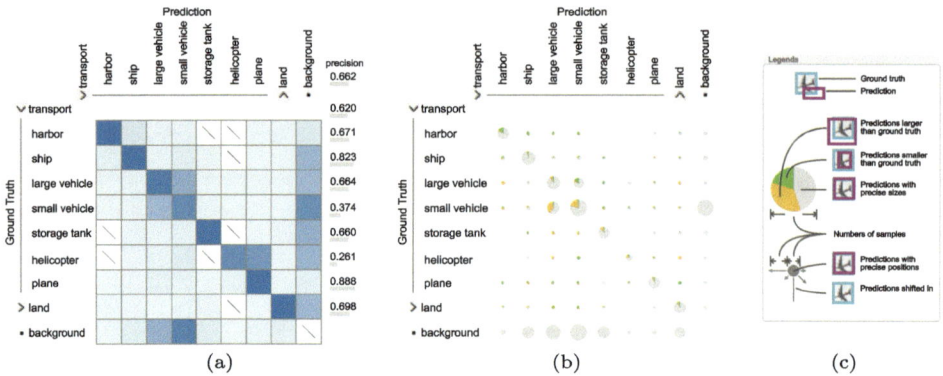

Fig. 4.15 Uni-Evaluator [41] provides a unified model diagnosis method for multiple computer vision tasks: **a** the confusion mode for evaluating classification performance; **b** the size mode for analyzing the sizes of predicted objects; **c** the legend for size mode and direction mode

ber of predicted objects in this sector, and the radius of the pie chart encodes the number of predicted objects in that cell. The direction mode utilizes eight arrows to encode the shifted directions of predicted objects. The length of each arrow encodes the number of predicted objects shifted in that direction (Fig. 4.15c). The circle in the middle represents the predicted objects with precise positions, whose radius encodes the number of such objects.

In addition to Blocks and Uni-Evaluator, there are many similar VIS4AI examples that extend confusion matrix visualizations, such as Neo [81], ComDia+ [192], and Confusion-Flow [95]. Moreover, matrix visualizations are widely used to illustrate the causal relationship between input data features and output predictions of ML models. Notable examples in this category include RuleMatrix [173], ExMatrix [181], and semantic navigator [110].

4.2.3 Tree Visualizations

Tree visualizations are commonly used to depict the hierarchical structure in ML data. The two most prevalent methods for tree visualizations are node-link diagrams and treemaps. A node-link diagram uses nodes and links to illustrate the tree nodes and the parent-child relationships between them. Different visual channels of nodes and links can encode additional attributes of the data. On the other hand, a treemap represents hierarchical data using nested rectangles. Each branch of the tree is represented by a rectangle, which is then tiled with smaller rectangles representing sub-branches. The size and color of each rectangle can vary, typically proportional to a specific quantitative attribute of the data. Both techniques have been widely used to visualize large volumes of data instances involved in ML models, such as training instances and neurons inside DNNs.

Liu et al. [156] developed a visual analytics tool, BOOSTVis, to understand and diagnose tree boosting models. The primary challenge is to create visualizations for a large number of decision trees, with the intent of displaying the variety of tree structures, the size of the trees, and the distribution of instances inside individual trees. To achieve these goals, a large number of decision trees are clustered based on their structure similarity. The adopted clustering algorithm is k-medoids, and the structure similarity between trees is measured through an extended version of the tree edit distance. As shown in Fig. 4.16a, the tree cluster component provides an overview of the tree structures through five clusters. Inside each cluster, the most representative tree is visualized as a node-link diagram. As this tree can have many nodes, a tree cut algorithm is adopted to highlight the nodes of interest and gray out the others. The histogram in Fig. 4.16b shows the changes in tree size during model training. The horizontal axis denotes the training iterations over time, and each bar denotes the number of tree nodes in an iteration. When a tree cluster is selected in Fig. 4.16a, the corresponding bars will be highlighted in Fig. 4.16b. Users can interact with these bars to select a specific tree in which they are interested. The decision tree component of BOOSTVis (Fig. 4.16c) then presents the instance distribution within the tree and the features used to split the set of instances. Specifically, the label on the internal tree node shows the feature name and the corresponding split value. The leaf nodes display their gradient values. The distribution of the instances on the tree is encoded by using the edge thickness, which is proportional to the number of instances that flow through this edge. In this component, the selected instances are represented by using the color of their actual class while others are colored gray.

Adversarial attack is a prevalent topic in the field of deep learning. In the context of image classification models, such attacks introduce imperceptible noise into an image. This can fool deep learning models into making incorrect predictions. This vulnerability poses significant challenges to the security and reliability of AI systems, particularly in sensitive applications such as autonomous vehicles, facial recognition, and cybersecurity. To diagnose how adversarial attacks work on a CNN, Cao et al. [27] introduced AEVis to perform multilevel visual analysis on three images: (1) a normal source image, (2) its adversarial counterpart, and (3) a normal target image. For example, if a "panda" image is mispredicted as a "monkey" due to adversarial attacks, the three images will be the original "panda" image, the "panda" image with imperceptible noise, and a normal "monkey" image. Through comprehensive network-level analysis, users first locate the layers where the activations of the adversarial image diverge from the original image or merge into the target image. These layers are responsible for the misclassification of adversarial examples. To investigate the key features learned by these layers, the authors conduct a layer-level analysis. They utilize a treemap to compare the feature maps of the three images within a layer, denoted as sets A, B, and C in Fig. 4.17a. The process begins by computing the shared and unique parts of the

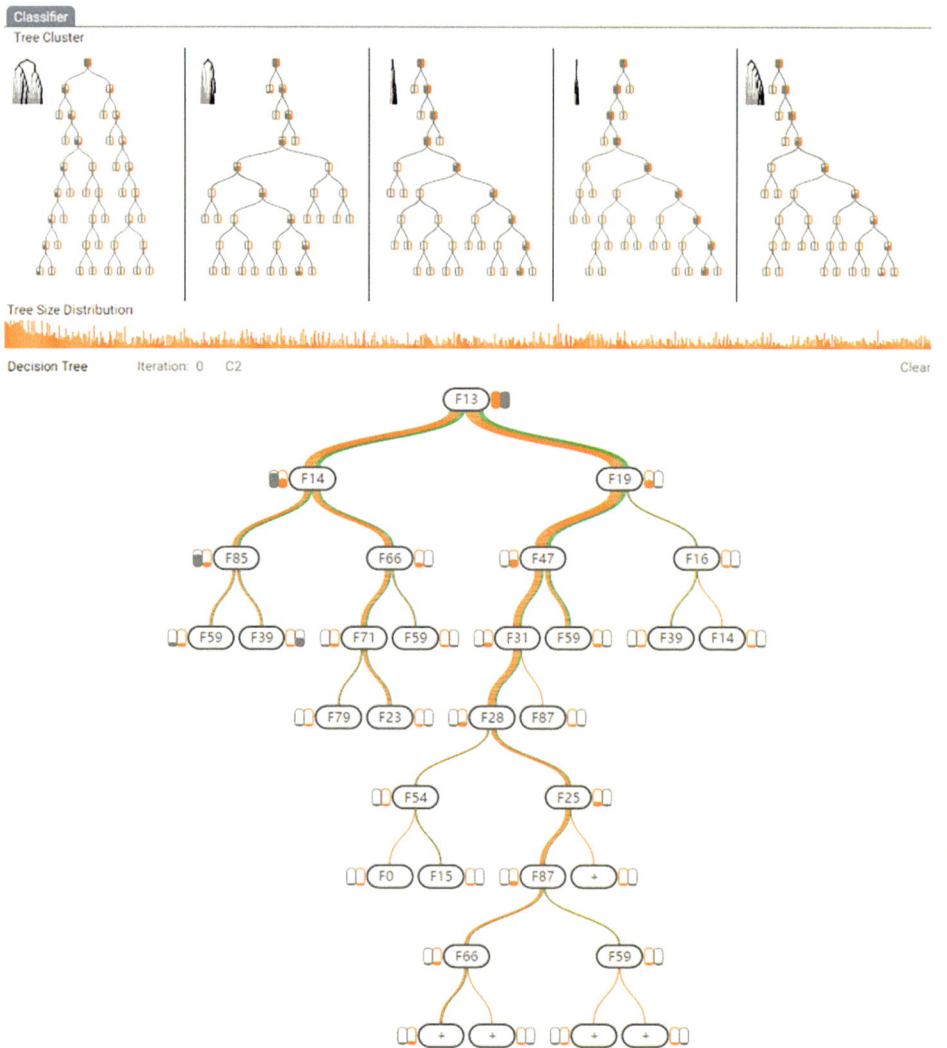

Fig. 4.16 Diagnosing tree boosting models with BOOSTVis [156]: **a** dividing a large number of decisions trees into five clusters; **b** visualizing the tree sizes with a bar chart; **c** displaying the feature splitting values in the internal nodes and encoding instance distributions with the edge thickness

three sets (Fig. 4.17b). Then, a hierarchy is built according to their set inclusion relationships (Fig. 4.17c). For example, $A \cap B \cap C$ is included in $A \cap B$, and thus its corresponding node in the hierarchy is a child of the node corresponding to $A \cap B$. Finally, a squarified treemap is generated to visualize this hierarchy (Fig. 4.17d), where the nodes with shared feature maps are placed in the center, and the nodes with unique feature maps are placed on the boundary. Feature maps in the treemap are visualized as squares with green filling patterns, where their

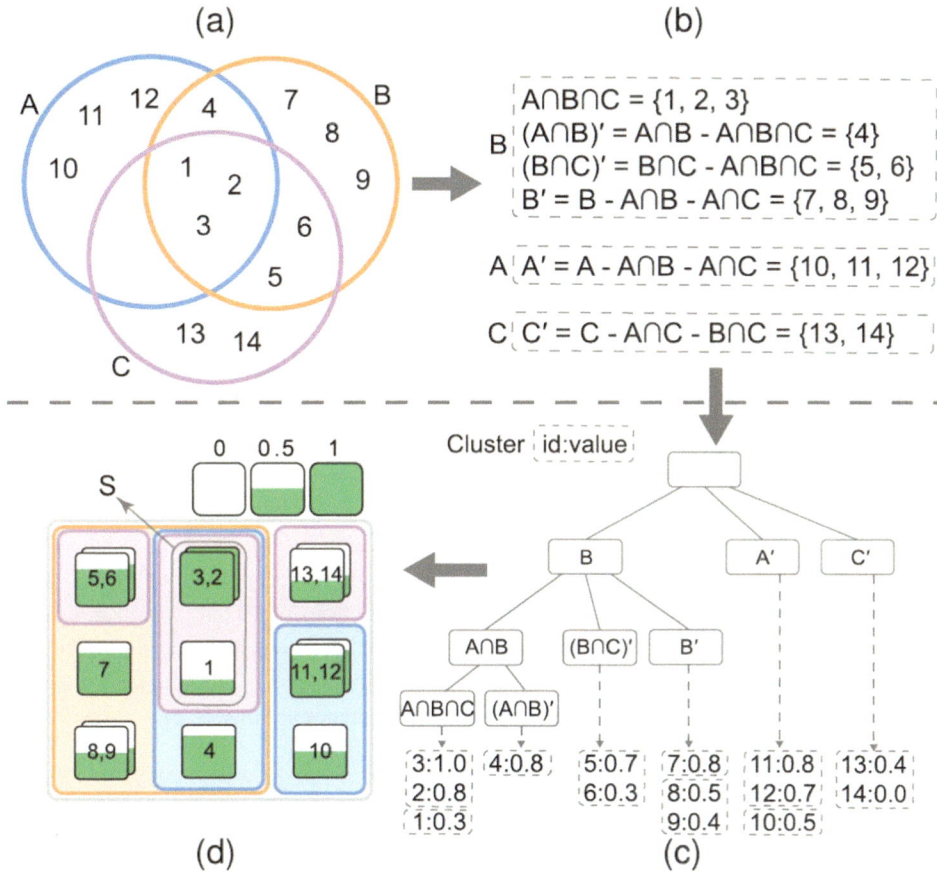

Fig. 4.17 AEVis [27] employs treemaps to explore the feature maps from a CNN layer: **a** three sets of feature maps (denoted as A, B, and C) from three images of interests are identified; **b** deriving the inclusion relationships among the three sets; **c** a hierarchy is built on the relationships; **d** visualizing the relationships as a treemap following the hierarchy

height encodes activation differences or contributions. Such hierarchical exploration helps locate the most salient feature maps and the corresponding neurons. Users can then drill down to conduct the neuron-level analysis on them.

There are more VIS4AI studies adopting tree visualizations for hierarchical data presentations. For example, both Baobabview [69] and VISTB [243] utilize node-link diagrams to visualize decision trees. The interactions provided by these visual analytics tools help users better explore and refine the tree structures. DeepCompare [180] organizes the test data instances into a hierarchy based on their ground-truth label and their prediction correctness from two ML models. The hierarchy is then visualized as a treemap to compare the prediction agreements and disagreements between the two models. More visual analytics tools with tree visualizations also include GBRTVis [104], CNN2DT [109], DendroMap [18], and Seq2Seq-Vis [228].

4.2.4 Sankey Diagrams and Parallel Sets

Sankey diagrams and parallel sets are similar visualizations that depict how data instances are distributed within a dataset. Their key strength lies in their ability to compare the distributions between different data features and to track the evolution of the distributions. These visualizations have been widely used to present the distributions of input data or the output predictions of ML models. Often, they facilitate the identification of specific instance groups, which subsequently form the basis for model diagnosis.

Wang et al. [243] introduced a visual analytics tool to investigate the evolution of tree boosting models. Its essence is to track the output predictions of all data instances across a large number of tree boosting iterations and diagnose the instances with deteriorated performance. At any iteration, data instances can be classified as correctly predicted or incorrectly predicted. They are placed into the diagonal and non-diagonal cells of the corresponding confusion matrix. Between iterations, the prediction of an instance can be "improved," "degenerated," "shifted," or "unchanged." As illustrated in Fig. 4.18a, instances demonstrating improvement are those whose predictions transition from incorrect (at iteration T_1) to correct (at iteration T_2). Conversely, degenerated instances are those whose predictions change from correct (at T_1) to incorrect (at T_2). Shifted instances are those that remain incorrectly predicted across both iterations but have transitioned from being misclassified into one class to being misclassified into another. Unchanged instances maintain the same predictions across both iterations. Accordingly, a Sankey-diagram-based temporal confusion matrix is designed to monitor the evolution of predictions. As shown in Fig. 4.18b, the training process contains 400 iterations, with confusion matrices in five key iterations, 1, 100, 200, 300, and 400, selected for visualization. Each column of Sankey nodes vertically stacks the non-empty cells of the corresponding confusion matrix. A Sankey node is named by $T\{iteration\}\_\{true\ class\}\_\{predicted\ class\}$. For example, node $T1\_0\_1$ includes instances whose true class is 0 but are misclassified as class 1 at iteration T_1. Bands connecting adjacent Sankey nodes illustrate the flow of data instances between the corresponding confusion matrix cells. The color of the bands represents the instance type in the associated data streams (improved: green; degenerated: red; shifted: blue; unchanged: gray). By default, the height of the Sankey nodes and the width of the bands are proportional to the number of instances in the corresponding confusion matrix cells and data streams. However, given the potentially large variance in the number of instances within individual cells, the authors also

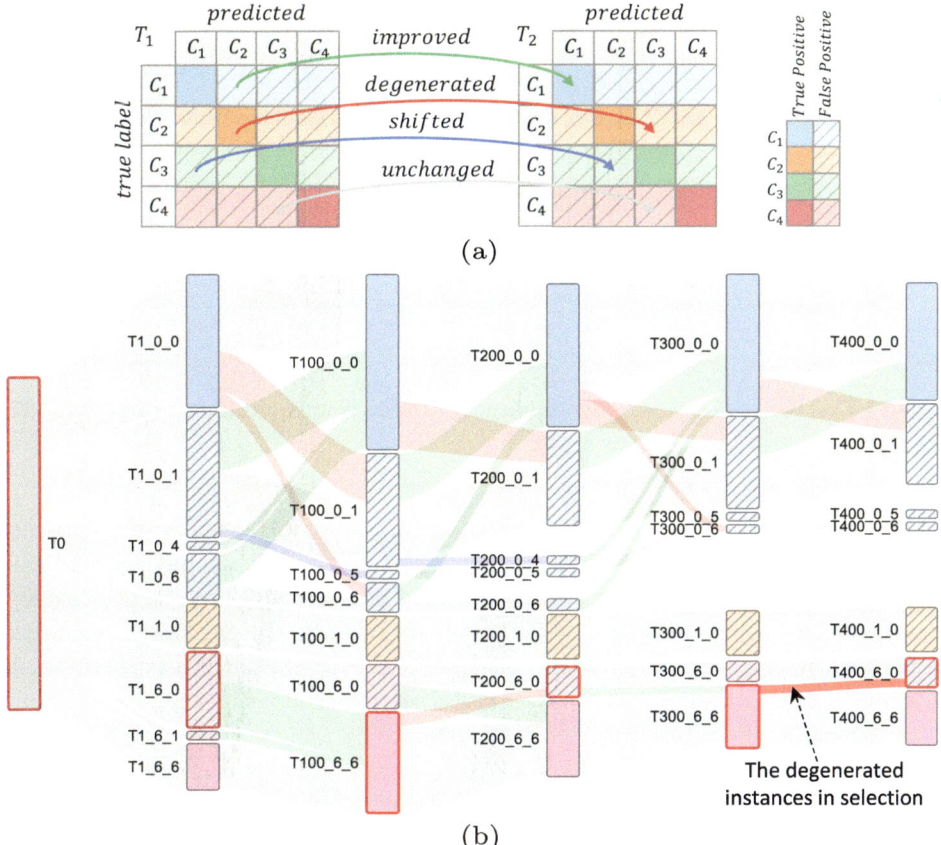

(a)

(b)

Fig. 4.18 The Sankey-diagram-based temporal confusion matrix: **a** data instances are categorized into four groups based on their prediction changes between two iterations; **b** a Sankey diagram is used to present the confusion matrices at multiple iterations and the prediction changes between iterations

implemented a log-scale mapping, which allocates more space to nodes/bands with fewer instances. In Fig. 4.18b, the red band between iterations 300 and 400 attracts more attention, as the predictions of these instances still degenerate in later iterations. To investigate the reasons behind this, users can directly click the red band to select these instances and use other views to further diagnose the model behavior.

Li et al. [145] developed DeepNLPVis to facilitate the understanding and diagnosis of NLP models in a unified way. This visual analytics tool employs a Sankey diagram to reveal how the model processes a word based on its context. The context of a word is represented by a list of words that are considered to be the most relevant to it according to the NLP model. The Sankey diagram then effectively illustrates the context of a word across different samples and model layers. Figure 4.19 shows an example Sankey diagram of the word "care." At

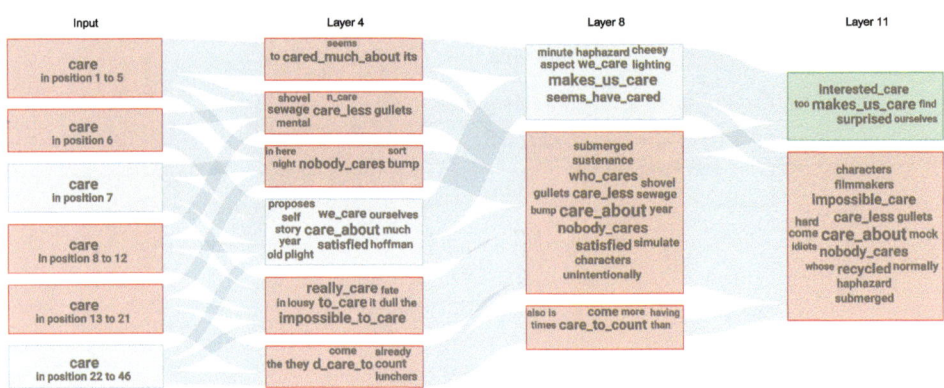

Fig. 4.19 DeepNLPVis [145] uses a Sankey diagram to reveal how the model processes a word based on its context

the beginning, the sample clusters are computed by performing agglomerative clustering on the feature vectors of the context words for "care." To maintain stability, the clusters in each layer are constrained by the clusters extracted in the previous layer. Each cluster is then represented using a rectangle in the Sankey diagram. Its size encodes the number of samples, and its color encodes the major polarity of "care" in these samples. For example, the meaning of "care" in the red cluster in layer 11 has negative sentiments (e.g., "nobody cares" and "care less"). The cluster positions are determined by using the directed acyclic graph layout algorithm employed in TextFlow [49]. The width of the edge encodes the proportion of samples that come from the previous cluster. By examining this Sankey diagram, users can analyze the word "care" in the context of relevant words and how the model processes it through layers.

In addition to the aforementioned examples, there are other VIS4AI studies toward adopting Sankey diagrams and parallel sets for model diagnosis. For example, InstanceFlow [199] uses a Sankey diagram to present the fraction of instances that flow between different classes. FairRankVis [262] employs parallel sets to depict the distribution of selected graph nodes across multiple protected features. ModelWise [169] uses Sankey diagrams to compare the performance of different ML models and diagnose the performance of different data subsets. DECE [43] utilizes Sankey diagrams to reflect the relationship between the distribution of the original input data of ML models and that of their counterfactual examples.

4.2.5   Customized Visualizations

The task of interpreting and diagnosing ML models often contains a range of unique requirements that require the use of customized visualizations to address these needs. This cus-
tomization process effectively leverages visualization capabilities and noticeably contributes
to VIS4AI. In addition, it is often necessary to integrate these customized visualizations
with other visualizations. This integration serves to provide coordinated and comprehensive
insights into the interpretation and diagnosis of ML models.

Cao et al. [27] introduced a customized DAG-based visualization in AEVis to diagnose
how adversarial examples successfully deceive a well-trained CNN classifier. In the image
domain, an adversarial example is an image generated by adding imperceptible noise to a
normal image of class A. This adversarial example can fool CNN classifiers to predict it as
class B ($A \neq B$). To diagnose this process, users investigate the internal CNN representations
of three images: (1) the original image of class A, (2) its adversarial counterpart, and (3)
an image of class B. A and B are "panda" and "monkey" in this work. The investigation is
carried out at three levels: network level, layer level, and neuron level (Fig. 4.20a).

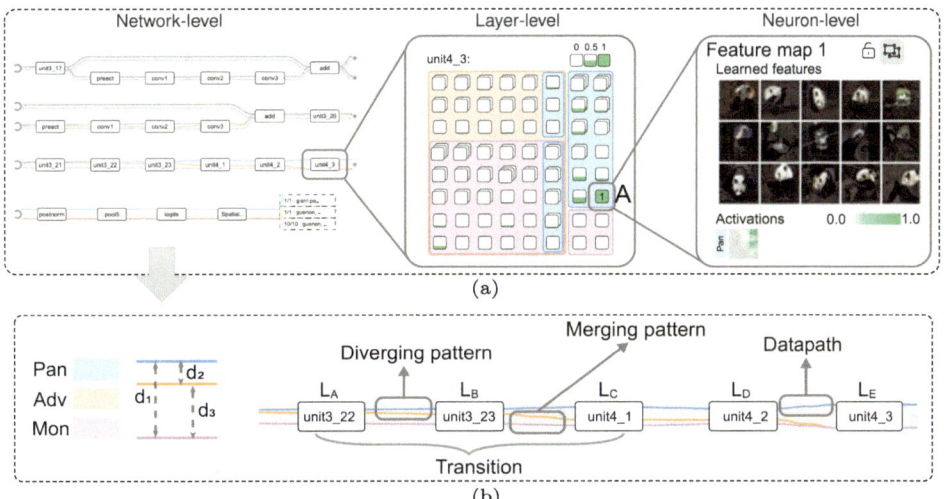

Fig. 4.20 AEVis [27]: **a** diagnosing how adversarial examples work at three levels: network, layer,
and neuron; **b** the river-based visualization facilitates the exploration of datapaths and helps locate
critical layers for further analysis

At the network level, the CNN architecture is formulated as a DAG, in which the nodes are layers, and the edges are the datapaths of the three images. The datapath of an image is defined as the critical neurons and their connections inside the CNN, which can be extracted by solving a subset selection problem. The DAG is then presented as the customized visualization in Fig. 4.20b, inspired by a river metaphor. The colors blue, orange, and purple represent the datapaths of the normal panda image, the adversarial panda image, and the normal monkey image, respectively. The distances from the orange curve to the blue and purple curves (denoted as d_2 and d_3) are proportional to the similarities between the corresponding datapaths. Following the three datapaths, it is straightforward to identify the layers where the datapath of the adversarial panda image diverges from that of the normal panda image and merges into that of the normal monkey image. Specifically, the divergence occurs between layers "unit3_22" and "unit3_23", while the merging occurs between layers "unit3_23" and "unit4_1". These transition layers are critical to diagnosing how the adversarial example works. The customized river-based visualization helps users locate them intuitively. Focusing on them, AEVis then presents layer-level details by organizing all neurons of a layer into a treemap visualization. This treemap guides users to further drill down to important neurons for more fine-grained neuron-level investigations. These details explain which neurons have been attacked and what features they have extracted, facilitating the diagnosis of adversarial attacks.

Seq2Seq-Vis [228] introduces customized visualizations to diagnose incorrect language translations generated by sequence-to-sequence models. A sequence-to-sequence model converts a source sequence to a target sequence over five main stages: (S1) encoding the source sequence into a sequence of hidden states; (S2) decoding the target sequence into a sequence of hidden states, with each state considering only its preceding tokens; (S3) attending the hidden states in the target sequence to those of the source sequence; (S4) predicting the next token in the target sequence; and (S5) searching for the best complete target sequence through beam search. Seq2Seq-Vis diagnoses a sequence-to-sequence model by allowing users to form hypotheses regarding errors originating from each of the five stages. Visual insights revealed from the combination of multiple customized visualizations help reject or accept those hypotheses one by one to locate the performance issue.

For example, Fig. 4.21 shows the process of diagnosing an incorrect translation in stages S3-S5. Stage S3 (Fig. 4.21a) uses a node-link-based visualization to depict the associated attention. A node represents a token in the source or target sequence, and a link between a source and a target token represents the attention between them. The width of the links encodes the strength of the attention. As shown by the thicker red link in Fig. 4.21a, the attention of the eighth target token is mostly distributed to the correct token, "dunkel," in the source sequence. This observation rejects the hypothesis that the error in translating "dark" to "to" comes from S3. For the prediction in stage S4, the predicted probabilities for possible words at each token position are visualized as a variant of the bar chart in Fig. 4.21b. The width of each bar encodes the probability of the corresponding word. In this case, the probabilities for the words "to" and "dark" are very similar. Therefore, the error

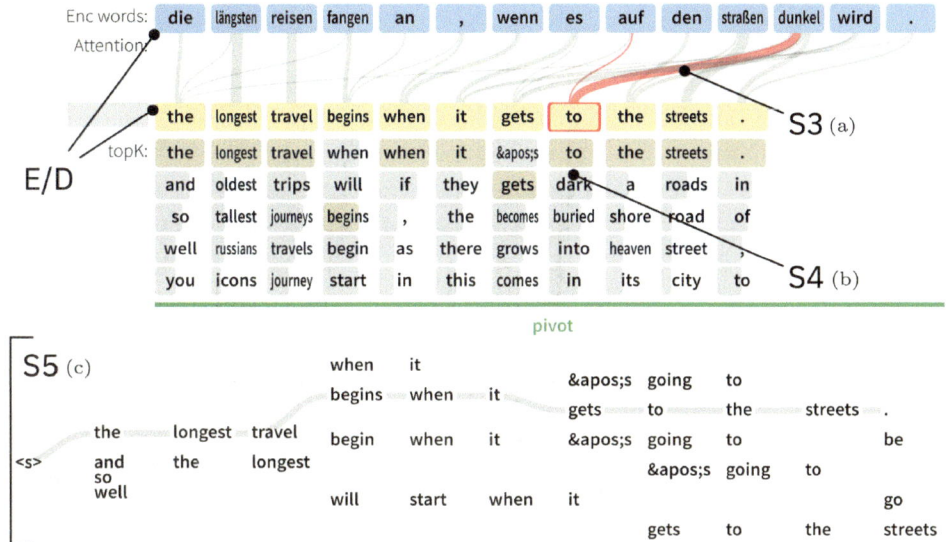

Fig. 4.21 Diagnosing sequence-to-sequence models with Seq2Seq-Vis [228]: **a** visualizing the attention between source and target tokens in stage S3; **b** encoding the prediction probability in stage S4 into a variant of the bar chart; **c** presenting the beam search tree in stage S5 as a node-link diagram

is also less likely to be caused by the prediction. Stage S5 determines the global optimal word at each target token position sequentially using the beam tree search method, which is visualized by a node-link diagram (Fig. 4.21c). Each node at each tree level represents a possible word at a specific token position. The strength of the connections between nodes at adjacent levels is indicated by the thickness of the connecting links. The thickest path in the node-link diagram shows that the model selects the word "to" at the eighth token position to optimize the global translation result. To probe the model and verify the hypothesis that the error comes from the beam search, the authors substitute the word "to" with "dark" at the eighth token position. This what-if analysis reveals the source of the error, which helps users identify the issues efficiently and then address them.

There are more works utilizing customized visualizations to present information in diverse formats and provide insights from various perspectives. For example, Wang et al. [246] diagnosed the discrimination in ML models with a customized set visualization named RippleSet. Ma et al. [164] designed a series of visual representations from overview to detail to reveal how data poisoning makes a model misclassify a specific data sample. By comparing the distributions of poisoned and normal training data, users can deduce the reason for the misclassification of the attacked sample. Additionally, many customized visualizations [6, 41, 81, 204] have been developed to better present the confusion matrix of classification models.

4.3   Model Steering

The interactive nature of visualization has made it possible to steer the training and execution of ML models. This includes model fine-tuning, selection, and ensembling, where human interventions can directly improve the typically automated process. This human-in-the-loop process often requires the coordination of several visualization techniques across multiple linked views. In this chapter, we take several visual analytics tools as examples to explain how ML models have been steered to achieve two major research goals: (1) model refinement and (2) model selection and ensembling.

4.3.1   Model Refinement

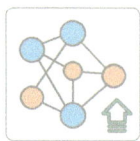

Model refinement is the process of tuning ML models to increase their accuracy, efficiency, or overall usefulness. Advanced visualization techniques enable a more intuitive integration of human insights and knowledge into this process. The refinement can be accomplished from various perspectives, such as improving the quality of training data, optimizing the model architecture, and adjusting hyperparameters.

Ming et al. [175] introduced a visual analytics tool, ProtoSteer, to interpret and steer deep sequence models. This tool is built on the inherent interpretability of a prototype sequence network (ProSeNet) [174]. Specifically, a ProSeNet uses an RNN encoder to transform an input sequence, such as an English sentence, into a latent representation. This latent representation is then represented as a set of prototypes through prototype learning. A prototype, a representative instance from the training data, carries semantics that support the interpretability of ProSeNet. The architecture of ProSeNet ends with a fully connected layer that learns the proper combination of prototypes based on the label of the input sequence. Well-trained ProSeNet models distinguish themselves not only through better performance but also through their enhanced interpretability. The semantics embodied by each prototype contribute significantly to the interpretability of the model. Thus, the quality of these prototypes is important. However, prototypes often contain redundant or less significant tokens, which hinders the effective learning of ProSeNet models and compromises their interpretability. Currently, there is no effective algorithm to simplify the prototypes. Developing such an algorithm is also challenging due to the complex and domain-specific rules that the input sequences must follow. Therefore, experts with domain knowledge are best suited to guide and improve the quality of prototypes.

ProtoSteer provides a visual interface that enables domain experts to add, delete, or revise prototypes. Figure 4.22 illustrates a use case where an NLP expert explores the ProSeNet

Fig. 4.22 Using ProtoSteer [175] to refine the prototypes: **a** two prototypes (#0 and #56) with very similar semantic patterns from the Yelp Reviews dataset; **b** removing prototype #0 and simplifying prototype #56

model trained on the Yelp Reviews dataset. The dataset contains millions of reviews, each of which is a sequence of words. The goal of the model is to predict whether a given review is positive or negative. In Fig. 4.22a, each prototype is represented by a row of words, with colors indicating their co-occurrence frequency with positive (blue) or negative (red) labels. The importance of each word in the prototype is represented by the width of the gray bar below it, which is computed by the leave-one-out ablation strategy. The bar to the right of each prototype shows the accumulated number of similar instances in close proximity to that prototype. By sorting all prototypes in order of their similarity and examining their latent representations encoded by the RNN encoder, the expert identifies two prototypes with very similar semantic patterns: prototype #0 and #56. A review of the raw text of these prototypes reveals that both express negative sentiments regarding the unsanitary environment or food. Including both prototypes in the collection is redundant. Therefore, the expert removes prototype #0, and in the meantime, simplifies prototype #56, as indicated by the red cross and blue exclamation mark icons beside the two prototypes in Fig. 4.22b. After fine-tuning the ProSeNet model with the new collection of simplified prototypes, the expert observes that instances that were originally closely associated with prototype #0 have transitioned to become closely associated with prototype #56. This transition is reflected by the two bars and the thick curve between them on the right of the two prototypes, confirming the success of merging them. By iteratively working with ProtoSteer, the expert successfully reduces the number of prototypes from 70 to 46, and decreases the average prototype length from 18.2 to 14.9. Meanwhile, the performance of the ProSeNet model remains consistent both before and after the prototype revisions. Throughout this process, ProtoSteer largely simplifies the prototypes and makes them more adaptable to the user specifications.

Yang et al. [264] developed ReVision to integrate both public knowledge from knowledge graphs and private knowledge from user preference into the hierarchical clustering of documents. After projecting documents into the relevant nodes in the knowledge graphs, an ant-colony-based algorithm [59] is employed to extract a constraint tree. This tree represents a coarse relationship between the documents and guides the fine-grained clustering. As shown in Fig. 4.23, the visual analytics tool contains two juxtaposed node-link diagrams

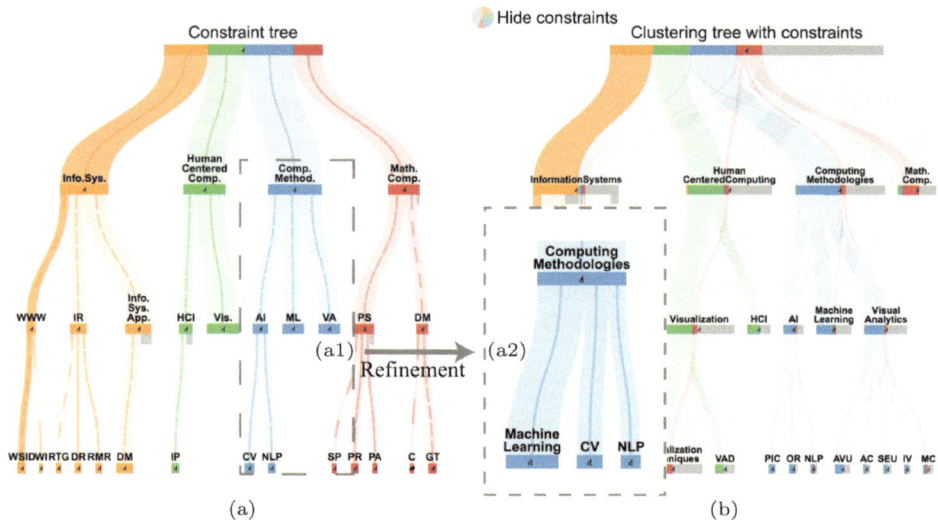

Fig. 4.23 ReVision [264] that steers the hierarchical clustering of documents: **a** the constraint tree; **b** the hierarchical clustering tree

for two tree structures: the constraint tree (Fig. 4.23a) and the hierarchical clustering tree (Fig. 4.23b). Each node represents a collection of documents, where the width encodes the number of documents, and the color encodes the document category based on the first level of the constraint tree. For those documents that are not in the constraint tree, their colors are set as gray. The link between a parent and a child node represents the sub-category relationship, and the dashed line in the center of each link encodes the uncertainty level of this relationship.

The two trees are coordinated to facilitate user explorations and help users inject their private knowledge to steer clustering results. When users click on a node from the constraint tree, the nodes in the hierarchical tree containing the corresponding documents will be highlighted. Users can interactively merge and remove nodes to reconstruct the hierarchy according to their domain knowledge. An example of this is shown in Fig. 4.23a1. In this case, the user observes that the "Visual Analytics (VA)" node in the blue branch is less related to its siblings, namely the "Artificial Intelligence (AI)" and "Machine Learning (ML)" nodes nested under the "Computing Methodologies" node. Instead, it is more related to the "Human-Centered Computing" node in the green branch. As a result, the user moves the "VA" node to the green branch. Also, he moves the "Computer Vision (CV)" and "Natural Language Process (NLP)" nodes to the same level with the "ML" node based on his research interests in these topics (Fig. 4.23a2). These user-specified constraints help generate better hierarchical clustering results.

There are more studies aimed at refining ML models through interactive steering [45, 69, 138, 147, 229, 261]. For example, van der Elzen and van Wijk [69] introduced BaobabView, a tool designed to assist users in iteratively constructing decision trees using their domain knowledge. Users can refine the decision tree through direct actions, such as growing, pruning, and optimizing internal nodes. The refined trees can then be evaluated using various visual representations. Liu et al. [147] developed MutualRanker, a visual analytics tool that employs an uncertainty-based mutual reinforcement graph model for extracting key blogs, users, and hashtags from microblog data. This tool visually presents the ranking results, uncertainties, and their propagation with the help of a composite visualization. Users can examine the most uncertain items in the graph and adjust their ranking scores. These adjustments are incrementally propagated across the graph to update the model. Strobelt et al. [229] presented GenNI, a generation negotiation interface for steering RNN-based text generation models. As these models often generate diverse texts that may not align with the target applications, GenNI employs a constraint graph to exert explicit control over the generation process. The interface engages users in a "Refine-Forecast" loop, which enables them to steer text generation by interactively specifying and refining the constraints. Li et al. [138] introduced CNNPruner, a visual analytics tool designed to optimize the architecture of CNNs. Based on the model statistics provided by the tool, such as loss fluctuation and recovery capability, users can directly prune less crucial neurons, determine pruning ratios, and fine-tune CNNs to compress the models without significantly sacrificing their predictive performance.

4.3.2 Model Selection and Ensembling

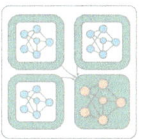

Model selection involves choosing the most suitable ML model from a set of candidates. Visualizations facilitate the understanding of the performance of each model across different data subsets and simplify their comparison and selection. Model ensembling, on the other hand, merges multiple models to enhance their collective performance. Visualizations clearly reveal the strengths and weaknesses of different models, which guides the ensembling process effectively.

Mühlbacher et al. [179] introduced TreePOD to steer the selection of decision tree models with different hyperparameter settings and varying model sensitivities. Specifically, Tree-POD first generates a large number of candidate decision trees by varying the hyperparameters such as the maximum tree depth, minimum leaf size, and maximum number of features. These trees are then presented as a scatterplot for an overview. As shown in Fig. 4.24a, each point represents a single decision tree, and the horizontal and vertical axes of the plot represent two metrics of the trees such as the number of nodes, tree depth, and model accuracy. In the context of model selection, not all trees in the plot are of equal importance.

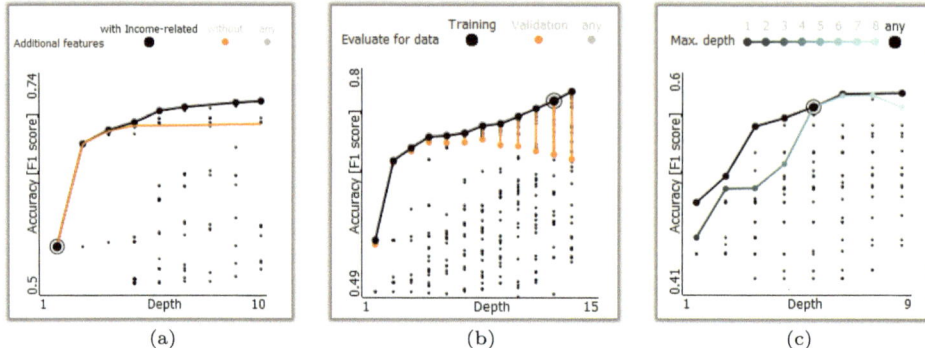

Fig. 4.24 TreePOD [179]: visualizing candidate decision tree models as points in the scatterplot and employing Pareto optimality to distinguish the important candidates as Pareto-front for interactive model selection

Therefore, the Pareto optimality [123] is employed to emphasize the most significant ones. A model from the candidate list is Pareto-optimal when no alternative model offers superior performance in certain criteria without compromising performance in other aspects. The set of Pareto-optimal models (according to the respective criteria) are highlighted with larger points in the scatterplot. Connecting them with a polyline forms the Pareto-front in the plot, and these decision trees emerge as stronger candidates for model selection. Varying the metrics represented by the two axes helps identify Pareto-optimal models with respect to the user-interested metrics.

The scatterplot of decision trees also supports the sensitivity analysis on different data features, different data subsets, and different hyperparameters. For example, in Fig. 4.24a, the orange polyline shows the Pareto-front of the tree candidates when excluding the "income-related" features during training. Deeper decision trees exhibit more severe performance degradation when trained without this feature. This illustrates the influence of this feature on the models. Figure 4.24b reveals that deeper tree models also experience more significant performance drops when transitioning evaluations from the training dataset to the validation dataset, which indicates potential overfitting issues of these deeper models. Figure 4.24c shows the performance of the selected Pareto-optimal model (the point in a circle) when varying its maximum tree depth. These immediate and interactive visualizations of the model metrics provide insights into sensitivity-aware model selection.

Das et al. [51] developed BEAMES for multi-model steering and selection in regression tasks. This visual analytics tool presents (1) a collection of regression models created using different algorithms along with their respective hyperparameters and (2) a data table that displays the training, test, or application datasets. The models are displayed as circular glyphs in a grid arrangement. Each glyph represents the performance of the corresponding model and indicates whether the model is recommended, liked, or selected by the user. From the data table, the user can interactively assign weights to individual data instances (rows of the data table) and select the features of importance (columns of the data table) based on his/her domain knowledge. Initially, the circular glyphs help users identify and select models that demonstrate superior overall performance. Users can then examine their effectiveness on critical data instances in the data table. This supports the exploration from models to data. Subsequently, focusing on these critical data instances, the tool recommends models that show enhanced performance specifically for these instances. This shifts the exploration from data to models. Users can increase the weight of these critical instances and emphasize certain features based on their domain knowledge. The models are then fine-tuned with these user inputs, and the tool is updated to reflect their new performance. Additionally, by evaluating the performance of the selected models across the training, test, and application datasets, users can further investigate and alleviate potential overfitting issues in the chosen models. This iterative exploration between the models and the data instances helps identify the optimal models that align most effectively with the critical data instances. Users can then save, like, or even export an ensemble of multiple models from the tool.

Yang et al. [265] developed FSLDiagnotor, which is designed to compare multiple base learners and interactively select an optimal combination to build a stronger ensemble model. It first utilizes pre-trained models to extract the features of each sample and then builds a set of learners based on the features. To facilitate the base learner selection, a sparse subset selection algorithm is applied to recommend an initial combination of base learners. However, the automatic selection results are not always perfect. To further improve the selection results, the learners and the selection results are presented in a matrix visualization. In this visualization, each row represents a base learner. The first column (Fig. 4.25A) encodes the number of samples predicted differently between a base learner and the ensemble model with a sequential color scheme. The darker the cell, the larger the difference. The remaining columns (Fig. 4.25B) present the comparison between a base learner and the ensemble model in terms of each class. A stacked bar is employed to visually convey the agreement and difference between the predictions. Specifically, the length of the stacked bar encodes the total number of the samples predicted to be of a certain class by either the base learner, the ensemble model, or both. To further facilitate the analysis of prediction differences, users are allowed to examine the confidence distribution by double-clicking on individual cells. A histogram will appear to present the confidence distribution of samples predicted by both the individual learner and the ensemble model (Fig. 4.25C). Through this visualization, users can understand the behavior of base learners and then make informed adjustments to the selection results.

Similar model steering works for model selection and ensembling are also commonly seen in the refinement of dimensionality reduction [62, 195], clustering [33, 189, 217], and regression models [21, 160, 197].

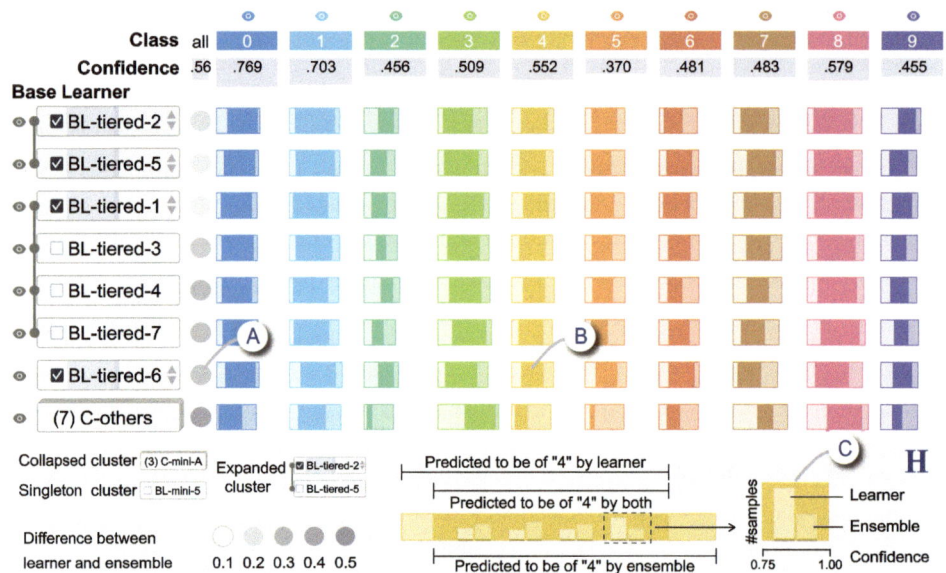

Fig. 4.25 FSLDiagnotor [265] employs matrix visualization with zoomable cells to disclose the prediction difference between base learners and the ensemble model

4.4   Summary

This chapter classifies the VIS4AI studies into three categories according to their tasks: model understanding, model diagnosis, and model steering. The first category enhances the understanding of ML models by revealing model details at various levels and correlating these details to offer insights into the models' inner workings. The second category focuses on the investigation of cases where ML models do not behave as expected and enables model developers to diagnose the reasons behind the unexpected behaviors. The third category integrates human input into the development of ML models through user-friendly interfaces and iterative user interactions.

Focusing on the prevalent visualization techniques employed in individual VIS4AI methods, we further elaborate on how these three tasks have been performed by exemplifying the methods. From the visualization perspective, most techniques serve more than one task. For example, node-link diagrams are utilized not only to understand the architecture of CNN models (Fig. 4.2), but also to diagnose the beam search component of sequence-to-sequence models (Fig. 4.21). From the task perspective, the success of a task often requires the support of multiple visualization techniques. For example, to diagnose the problematic part of a sequence-to-sequence model, multiple views in Fig. 4.21 work coordinately to enable users to examine individual model stages and conduct what-if analyses.

Table 4.1 Summary of the VIS4AI studies in the model development stage

Task	Papers
Model Understanding	[23, 29, 46, 63, 73, 76, 99, 107, 108, 114, 115, 129, 149, 158, 171–173, 180, 183, 203, 209] [216, 220, 227, 239, 241, 242, 256, 274, 279]
Model Diagnosis	[4, 6, 19, 26, 27, 55, 78, 91, 125, 150, 151, 156, 164, 196, 204, 224, 228, 240, 254, 275]
Model Steering	[21, 30, 32, 33, 45, 51, 56, 61, 64–66, 122, 128, 134, 147, 160, 166, 170, 175, 179, 189] [69, 197, 213, 217, 237, 248, 261, 264, 278]

There are many other VIS4AI studies that successfully accomplish the three tasks. We provide the following table for a concise summary. Table 4.1 provides a concise summary.

Techniques for Model Deployment

5

After preparing high-quality data and training the model on such data, the ML lifecycle reaches its final stage: deploying the trained model in a production environment. This stage poses significant challenges due to the diverse and heterogeneous nature of the production environment. As described by Hong et al. [102], a production environment typically involves a varied group of model consumers, including not only model developers but also domain experts, program managers, and business stakeholders. This diversity introduces complexity in achieving their goals, as the requirements of these consumers may differ significantly. Taking a resume screening model as an example, model developers primarily prioritize accuracy, while business stakeholders may value fairness more.

In the model development stage, the main focus has traditionally been on achieving higher accuracy. However, in the model deployment stage, there has been an increasing emphasis on transparency, trustworthiness, robustness, and fairness. These aspects are important for decision explanation, model monitoring, and model maintenance. **Decision explanation** illustrates how the model makes decisions, including individual predictions and the overall functioning of a model. This task is essential for ensuring transparency, fostering effective communication, and building trust among model consumers such as domain experts and business stakeholders. **Model monitoring and maintenance** focus on observing the model in use and making the necessary adjustments, which are critical for model developers and other consumers responsible for model performance. This includes maintaining robustness against changes in data over time and adversarial attacks, and ensuring the fairness of the model. Through continuous monitoring and maintenance, model developers can adapt the model to new data and conditions, thereby improving its performance. Figure 5.1 presents the key goals and tasks essential for deploying ML models, along with the visualization techniques that support these tasks.

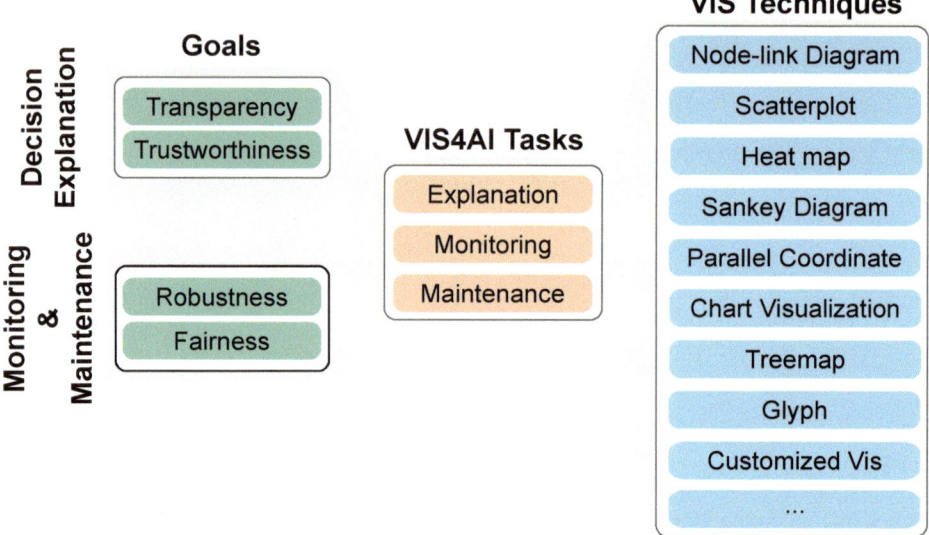

Fig. 5.1 An overview of the categories of goals, tasks, and visualization techniques for model deployment

5.1   Decision Explanation

Decision explanation discloses the rationale behind the final decision of a model and provides essential transparency to build user trust. In contrast to model explanations that explore the internal working mechanism of models, decision explanations focus on interpreting the final decision. As shown in Fig. 5.2, they are designed to be more straightforward and less technical. This makes them accessible and user-friendly even for those without experience in machine learning or model training. In addition, decision explanation methods are typically model-agnostic, which enables their application across different models and contexts.

There exists a variety of decision explanation methods, including local explanations that aim to derive detailed insights into individual predictions, as well as global explanations

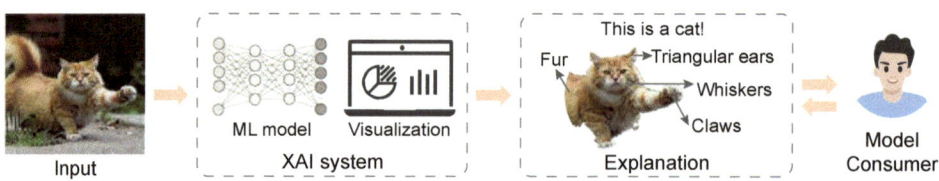

Fig. 5.2 Decision explanation helps model consumers understand the decision-making process of the model and thus enhances their trust in the model

that aim to understand the overall functioning of a model. Visualization serves as a valuable tool to enhance the understanding of both local and global explanations.

5.1.1   Local Explanations

Local explanation methods generally use a simplified and interpretable local model to approximate the prediction behavior of a more complex model in the neighborhood of a specific sample. One notable example is LIME [205], which employs a sparse linear model to generate local explanations. LIME first fits this local model over a sample of interest and its neighbors to produce predictions that closely resemble those of the original model. After extracting linear feature weights from the fitted local model, it generates visual aids, such as histograms or saliency maps, to visually illustrate the feature importance.

Perturbing the selected sample to generate its neighboring samples is an essential step in training the local model. The quality of these neighbors determines how well the local model approximates the behavior of the original model on these samples. However, naively altering a sample by deleting or modifying features can possibly lead to the creation of samples that lack semantic coherence such as fragmented sentences or images with holes in the middle. Such samples can introduce biases during the training of the local model, which may further reduce its interpretability. This challenge is particularly severe in scenarios where the input samples are images, as the nature of image data inherently complicates the generation of plausible and semantically meaningful perturbations.

To tackle this challenge, Wang et al. [241] developed DeepVID. This visualization tool generates semantically meaningful neighboring images by a DGM for a given image of interest. These generated images are fed into the original model to obtain their predictions. The images and their predictions are then used to train a linear local model to approximate the behavior of the original model on these images. As shown in Fig. 5.3, three coordinated views are designed to explain the behaviors of the original model (Fig. 5.3a), the VAE used to generate neighboring images (Fig. 5.3b), and the local model (Fig. 5.3c), respectively. For the original model, a scatterplot combined with t-SNE is used to show the distribution of all images (Fig. 5.3A$_1$). A confusion matrix is used to show the class confusion of the original model (Fig. 5.3A$_3$). By selecting the points in the scatterplot or clicking on a cell in the confusion matrix, the corresponding images are displayed in a grid visualization (Fig. 5.3A$_2$). Users can select an image of interest in the grid visualization to generate its local explanation and examine the predicted probability distribution by the original model with a bar chart (Fig. 5.3A$_4$). For the DGM, a parallel coordinate plot is employed to display the values of each dimension of the latent vector corresponding to the selected image (Fig. 5.3B$_1$). Users can specify the range of each dimension (the light blue band) to generate the neighboring images of the selected one, and the generated images are displayed in another grid visualization (Fig. 5.3B$_2$). For the local model, DeepVID also displays the predicted probability distribution of the selected image with a bar chart. Moreover, it generates a set of heat

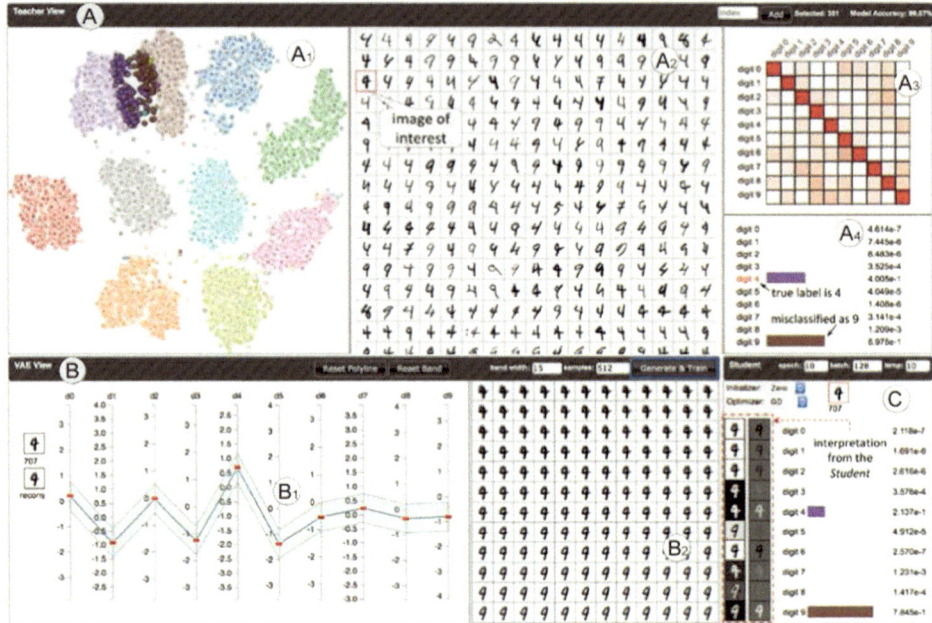

Fig. 5.3 DeepVID [241]: a visual analytics tool for understanding image classification models through local explanations

maps to illustrate how each pixel contributes to the prediction. Figure 5.3C takes the MNIST dataset as an example to illustrate the basic idea. Here, the user selects sample #707 (a digit "4") and examines its predicted probability distribution learned by the local model. A red dotted box encloses a heat map, with the second column representing the weighted features. Notably, white pixels in this column indicate a large contribution to the respective class. In this example, these white pixels exist in the heat maps of the classes "4" and "9," which illustrates the reason for the high prediction probability of these two classes.

There are more VIS4AI studies that support local explanations of models. For example, Krause et al. [125] proposed a method to generate local explanations for the prediction of individual samples in binary classifiers by probing the samples. They also developed several visualizations to explore these generated explanations by grouping samples with similar explanations together. Collaris et al. [46] developed ExplainExplore to help users explore and analyze the features relevant to model predictions with local models. To visually evaluate the accuracy of the local surrogate model, they designed a context visualization to display neighboring samples and the impact of perturbations on a selected sample.

5.1.2 Global Explanations

Decision rules are a common method to explain the overall behavior of a model globally. They show how the model reaches the final prediction for a group of samples that satisfy the condition specified by the rule [135]. However, visualizing decision rules can be complex, especially when the structure of a rule is highly complex or the number of rules is large. To solve this issue, Ming et al. developed RuleMatrix [173], which extracts a list of sequential decision rules to approximate the behavior of a classification model. RuleMatrix is a model-agnostic method that is applicable to explaining any classification model. As shown in Fig. 5.4B, a matrix is employed to visualize the extracted rules, where each row represents a rule, and each column represents a feature. The conditions of the features (e.g., $11.5 < alcohol < 12.3$ shown in the third rule) are visualized in the corresponding cells. Specifically, the distribution of a feature is visualized with bar charts (discrete features) or histograms (continuous features), and the group of samples that satisfy the condition is highlighted with an overlaid rectangle. To the left of the matrix, a waterfall-like Sankey diagram provides an intuitive representation of how samples are classified through this sequential rule list. The vertical branch represents the unclassified samples, and each horizontal branch represents the samples classified by the rule encoded by a row. The width of a branch indicates the number of samples, and the color represents their ground-truth labels. There are three additional columns to the right of the matrix that provide detailed information for each rule. The first column shows the prediction of each rule, where the color represents the label, and the number is the probability. A stacked bar adjacent to the colored number

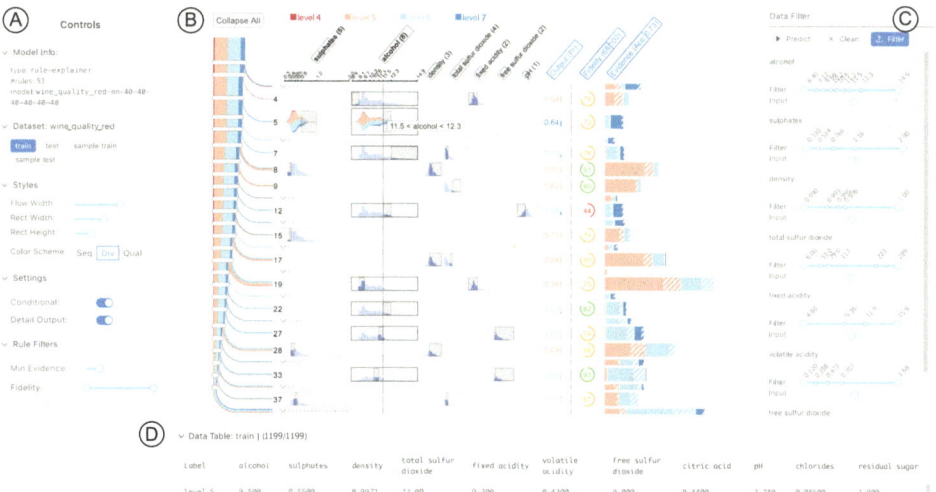

Fig. 5.4 The visualization of RuleMatrix [173], a visual analytics tool for explaining a model with sequential rules

shows the probability distribution for each label. The second column shows the fidelity of each rule, which means the consistency of its predictions with that of the original model on the samples covered by this rule. The last column employs stacked bars to illustrate the samples and their predictions made by each rule. The color represents the predicted label, and the solid and striped patterns indicate the correct and wrong predictions, respectively. Additionally, the length of the bars represents the number of samples.

Later work also employs this matrix visualization to visualize decision rules. For example, Neto et al. [181] developed Explainable Matrix to support global explanations of random forest models. Instead of fitting a surrogate model, it directly extracts decision paths from the decision trees in a random forest model and converts them into logic rules. A decision path refers to a path from the root node to a leaf node in a decision tree, and the corresponding logic rule combines all the conditions on this decision path. These rules are visualized in a matrix visualization. As shown in Fig. 5.5, each row represents a rule, each column

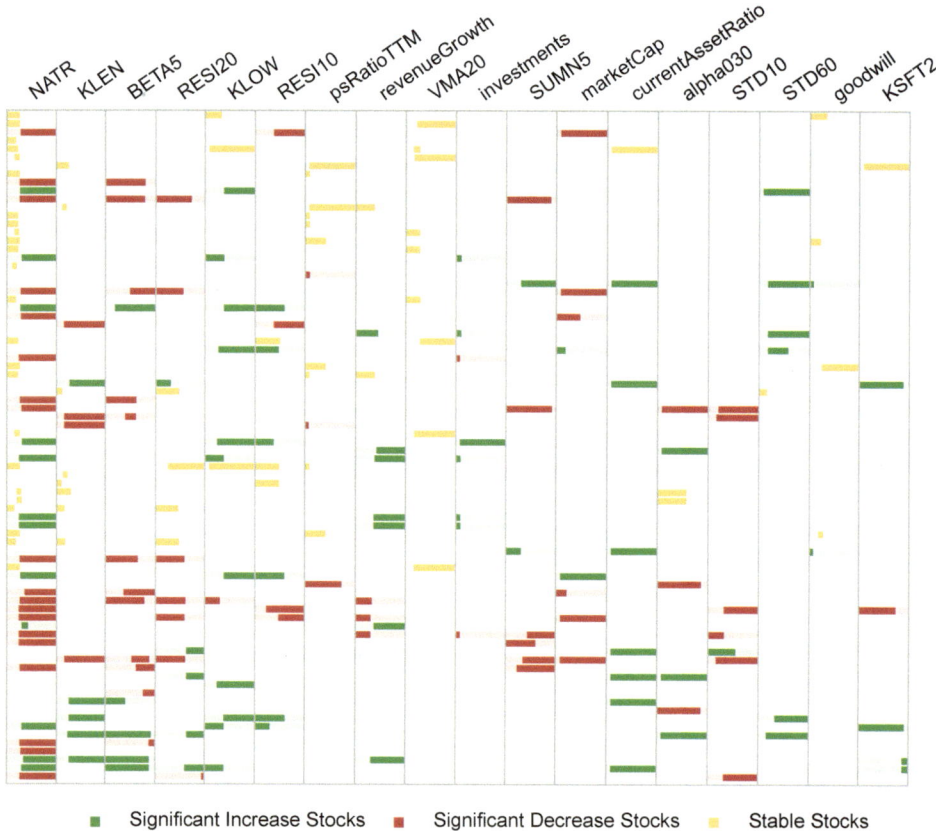

Fig. 5.5 Using the matrix metaphor to visualize a large number of rules extracted from a random forest model trained on a stock trading dataset

represents a feature, and each cell represents a rule condition. The colored bars in the cells represent the ranges of the feature values, where the colors encode the predicted labels of the corresponding rule. This simplified representation of rule conditions offers a more intuitive way to display rules compared to traditional tree-structured metaphors.

There are more VIS4AI studies that offer global explanations for model decisions. For example, Zhao et al. [279] developed iForest, a visual analytics tool that supports the interpretation and comparison of decision paths in random forests. In addition, in domain-specific applications, such as risk assessment, Migut et al. [171] proposed an interactive framework to explain the decision boundaries and performance of classification models through visualization. It helps domain experts explore the trade-offs between the classifiers and select the best decision point.

5.2   Model Monitoring and Maintenance

Model monitoring and maintenance involve observing the performance of the deployed model and updating it when needed. These two tasks are important for both model developers and other stakeholders responsible for ensuring optimal model performance. With the increased diversity in model consumers during model deployment, it becomes essential to extend the goals considered beyond mere accuracy. Factors such as robustness and fairness, which may not be apparent during model development, can surface during model deployment. Consequently, the monitoring and maintenance process involves (1) ensuring **robustness** against concept drifts and adversarial attacks; (2) guaranteeing **fairness** in model predictions and decisions.

5.2.1   Robustness

Ensuring the robustness of ML models is a key goal during model deployment. Developed models may experience a decline in accuracy in dynamic environments if they are not robust enough to handle unforeseen situations which were not considered during the model development stage. Typically, the poor robustness is caused by two main types of performance degradation: the emergence of concept drifts and adversarial attacks from malicious users. *Concept Drift*. In a production environment, the performance of ML models trained on historical data can significantly decrease due to unforeseen changes in future data distribution. These unexpected changes arise from two aspects: the distribution of input data and the relationship between the input and output data. Next, we will introduce how to effectively handle these two kinds of concept drifts with visualization techniques.

The first type of concept drift refers to the phenomenon that input distributions differ between the deployment and development stages. For example, a rainfall forecast model trained on historical data distribution might encounter a performance drop since the average temperature has steadily increased over time. Yang et al. [263] developed DriftVis to handle this type of concept drift in streaming data. First, a drift detection method is developed to calculate the drift degree based on data distribution. Without any prior assumption on the data, the difference between two distributions is measured with their energy distance. A larger difference between the historical and the incoming data indicates a larger drift degree. In addition, directly comparing the distribution between the incoming data and the entire historical training data may not be appropriate because the incoming data might be covered by a small subset of the training data. In this case, the performance of the developed model on the incoming data shows negligible change, but the calculated drift degree can still be large. To mitigate this problem, the drift degree computed for the incoming data should only consider the historical training data that is similar to them. Therefore, DriftVis clusters the streaming data based on the incremental Gaussian mixture model (GMM) [70] and computes the drift degree as the weighted sum of the energy distance for each cluster.

Then, they developed a stream-level and a prediction-level visualization to support the analysis of when, where, and why concept drifts happen. In the stream-level visualization, a line chart (Fig. 5.6a) is employed to visualize the computed drift degree and its temporal evolution. This visualization choice allows for straightforward identification of the specific time point at which concept drift occurs. In addition to the drift degree, the data distribution also needs to be visualized to reveal where and why the concept drifts occur. This is empowered by a novel dimensionality reduction technique, GMM-based constrained t-SNE, that projects

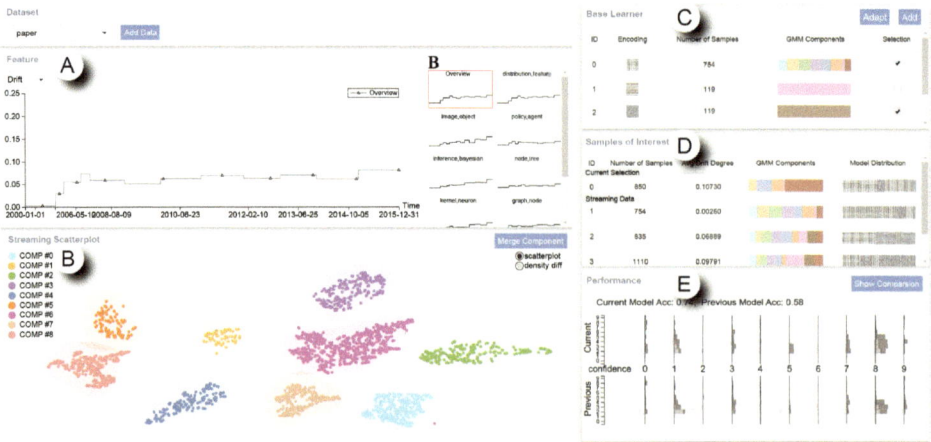

Fig. 5.6 The visualization of DriftVis [263], a visual analytics tool for analyzing concept drifts

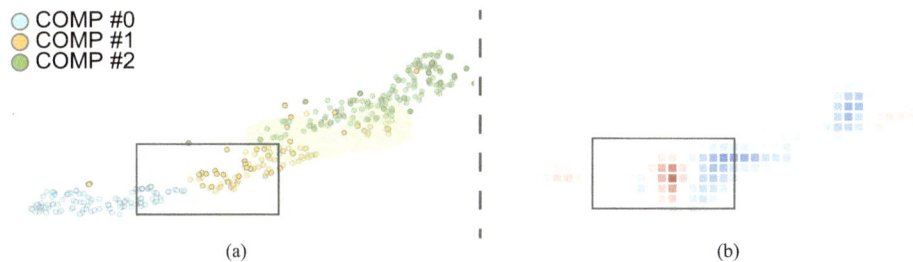

Fig. 5.7 Comparing the data distribution in DriftVis [263]: **a** representing the historical data distribution with a density map and the incoming data with scattered points; **b** representing the density difference with a heat map

the high-dimensional streaming data onto the 2D plane (Fig. 5.6b). This visualization helps explain the distribution of each Gaussian component in the GMM while maintaining the stability between the previous projection results and the new projection results with the incoming data. With the projection results, DriftVis generates two visualizations to show how the incoming data forms new components or deviates from existing ones. As shown in Fig. 5.7a, to compare the difference between historical data and incoming data, the historical data distribution is represented as a density map, and the incoming data is shown as scattered points. A multi-step animation is designed to convey how the data distribution is updated when new data flows in. Users can also select two specific data batches for comparison. As shown in Fig. 5.7b, a heat map is utilized to intuitively compare their density. The density difference is encoded using a dual color scheme, where red and blue indicate an increase and decrease of density, respectively, and a darker color indicates a larger difference.

After detecting and understanding the concept drifts, the prediction-level visualization is used to analyze model performance and adapt to drifts. DriftVis employs an ensemble method for drift adaptation due to its efficiency and flexibility. Correspondingly, two list views are utilized to present the information of base learners (Fig. 5.6c) and samples of interest (Fig. 5.6d), which helps users determine the optimal combination of base learners to form the ensemble model. Moreover, DriftVis utilizes the stacked bar chart design of Squares [204] to diagnose model performance. A side-by-side comparison is enabled on demand to verify model performance improvement after drift adaptation (Fig. 5.6e).

Some other work utilizes simple visualizations to explain concept drifts [55, 185, 191]. For example, Olson et al. [185] focused on detecting concept drifts in an image classification task. They employed a density ratio estimation algorithm to compute an outlier score for each test image. Subsequently, a side-by-side histogram visualization is utilized to compare the training and test images by grouping them based on their outlier scores, which facilitates the identification of concept drifts. Palmeiro et al. [191] studied concept drifts in fraud detection. They employed a heat map to visualize how much an input feature deviates from its historical distribution. In the heat map, each row represents a feature, each column represents a time slice, and a darker cell indicates that the corresponding feature is more likely to have a concept drift at the corresponding time slice. They also used a bar chart to display the distribution of the selected feature within a chosen time range and used lines overlaid on the bars to show its historical distribution. This design facilitates the comparison of feature distributions and the analysis of potential concept drifts.

The second type of concept drift refers to the phenomenon that the relationship between input and output data in the deployment stage differs from that observed in the development stage. For example, in the context of rainfall forecasts, the greenhouse effect has become more severe with its increasing effect on rainfall over the past years. As a result, the rainfall forecast model trained on historical data may not be appropriate for incoming data. Unlike the previous type of concept drift, which only involves changes in input data distributions, detecting and diagnosing the changes in the relationship between input and output data is a more complicated task. Besides, this type of concept drift also appears in multi-sourced data, where various sources exhibit similar and different concept drift patterns. Facing such problems, Wang et al. [249] developed ConceptExplorer, a visual analytics tool to compare and analyze concept drifts from multiple data sources. ConceptExplorer detects the accuracy drop of the prediction model to find potential concept drifts. Specifically, by segmenting data into uniform batches, the concept drift level of the current data batch is measured by how much the error rate on this batch is larger than the minimum error rate on the previous batches. Besides, they introduced a consistency judgment model to infer whether the concept drift in one data source occurs consistently with those from other data sources.

The analysis of concept drifts in multi-sourced data is supported by a set of coordinated visualizations, as shown in Fig. 5.8. After loading the dataset (Fig. 5.8a), a timeline visualization is presented for identifying the occurrence patterns of the concept drifts, where each data source is represented as a row (Fig. 5.8b). Marks in the rows indicate when concept drifts happen as well as their corresponding drift levels. After selecting a time period of interest, a prediction model view is shown to facilitate the analysis and comparison of the concept drifts in multiple data sources (Fig. 5.8c). In this view, line charts are utilized to show the accuracy of the prediction models along with the magnitude of accuracy drops when concept drifts are detected. A scatterplot shows the projection of the prediction model parameters of different data sources over time, and the prediction model parameters from the same data source are connected in temporal order. This eases the comparison of the parameter drifts between the prediction models of different data sources. In addition, a matrix visualization is employed to present the correlation between different data attributes, which helps explain the cause of concept drifts (Fig. 5.8e).

Adversarial Attack. In machine learning, various adversarial attacks have been proposed to intentionally mislead ML models to make incorrect predictions. These attacks pose a threat to the deployment of these models in real-world applications, especially in safety-critical scenarios such as autonomous driving and intelligent health care. It is therefore practically demanding to understand and address these issues for improved model robustness. Typically, the attacks take two forms: data poisoning and adversarial samples.

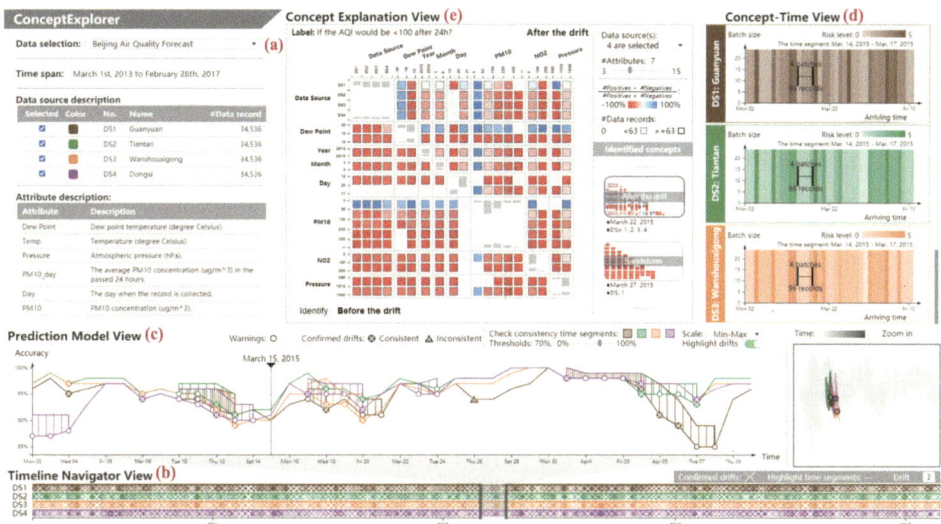

Fig. 5.8 The visualization of ConceptExplorer [249], a visual analytics tool for analyzing concept drifts in multi-sourced data: **a** the data entrance describes the loaded data; **b** the time navigator view showing the overview of the concept drifts; **c** the prediction model view showing the accuracy fluctuation and parameter changes in the prediction models; **d** the concept-time view showing the time segments being analyzed; **e** the concept explanation view used for analyzing the correlation between different data attributes

Data poisoning occurs when attackers manipulate training data from potentially untrustworthy sources. This often leads to incorrect model behavior on targeted test samples. For example, a spam filter can flip specific malicious emails from spam to non-spam by adding carefully crafted emails to the training data. These carefully crafted emails are called poisoning samples, which fool the model into mispredicting specific samples. To understand the impact of such attacks, Ma et al. [164] designed coordinated visualizations to identify robustness issues caused by data poisoning from multiple perspectives. As shown in Fig. 5.9, a radar chart compares the original and attacked model in terms of multiple performance measures such as true positive (TP), false positive (FP), and accuracy (Fig. 5.9c), a scatterplot provides an overview of the data (Fig. 5.9d), and several tables facilitate the detailed analysis of relevant samples and features (Fig. 5.9b, e, f). To understand how an attack impacts the model behavior, a local impact view is designed to disclose the neighboring relationships between samples and reveal how the poisoning samples affect the prediction of their neighboring samples (Fig. 5.9G). To reduce visual clutter, this view only shows three types of critical samples related to the attack: the samples being attacked, the poisoning samples, and the samples with flipped labels after the attack. Then, the k-NN graph of these critical samples is extracted and visualized with a node-link diagram. In this diagram, the poisoning samples and the k-NN samples are represented as solid circles, while the samples being

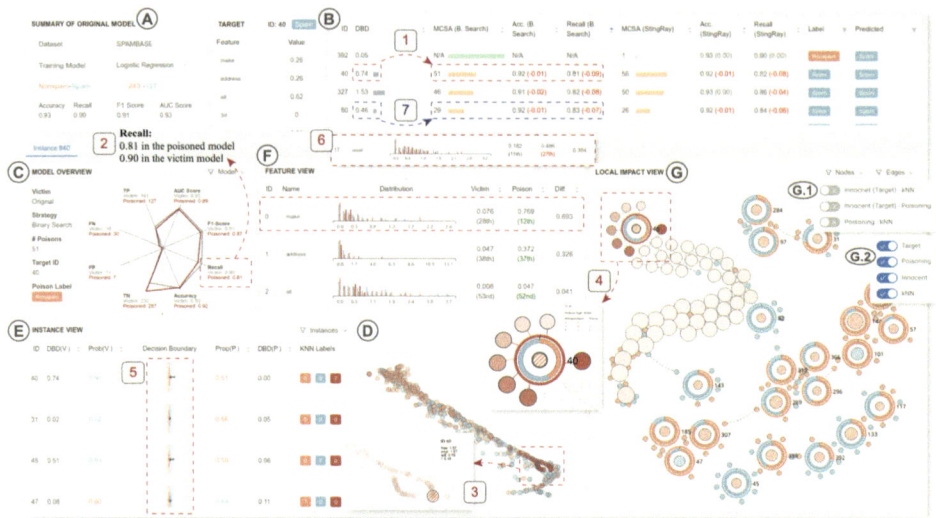

Fig. 5.9 Understanding data poisoning attacks with visual analytics

attacked and the samples whose labels are flipped after the attack are represented as nested glyphs. In this glyph, the inside circle uses color, texture, and size to encode the predicted label, whether the predicted label is flipped in the attacked model, and the predicted probability, respectively. The inner and outer rings represent the predicted class distribution of the original and attacked model on the k-NN samples, respectively. By exploring the neighborhood structures of these critical samples, users understand and analyze how they are impacted by data poisoning.

Adversarial samples are generated by adding imperceptible noises to normal samples. Although these adversarial samples are similar to the original samples, the model will mispredict them as another class. In the context of convolutional neural networks, Cao et al. [27] developed AEVis to explain the reasons behind such mispredictions by comparing the datapaths between normal and adversarial samples. The datapath of a sample refers to the neurons that highly contribute to the model prediction as well as their connections. Similarly, Das et al. [50] developed Bluff, a visual analytics tool that visualizes the activation pathways utilized by the samples. Slightly different from the datapath, the activation pathway of a sample refers to the neurons highly activated by the sample and the connections between these neurons. After extracting the activation pathways, Bluff utilizes directed acyclic graphs to visualize and compare the neurons activated by the normal and adversarial samples, explaining the reasons for the mispredictions.

5.2.2   Fairness

When using ML models for decision-making, it is crucial to prioritize fairness along with accuracy. Generally, a fair ML model treats specific features equally when making predictions. For example, an ML model used for college admission should be fair to the applicants in terms of gender. In real-world applications, the unfairness in ML models comes from data-related and model-related issues [4].

Data-Related Fairness. Several VIS4AI studies focus on identifying fairness issues caused by training data. Existing work primarily reveals fairness issues by visualizing three types of information: performance difference between sample subgroups, distortion of pairwise distances between samples, and counterfactual explanations.

One typical way to detect fairness issues is to investigate the performance of different sample subgroups and summarize the cases where the model produces unfair prediction results. Along this line, Cabrera et al. [26] developed FairVis to support the exploration of different sample subgroups and compare their performance, which helps users figure out where and why the unfairness exists (Fig. 5.10). It utilizes bar charts to provide a comprehensive overview and enables the comparison between different subgroups of data from multiple aspects, which helps determine the existence and reasons for unfairness. Users can

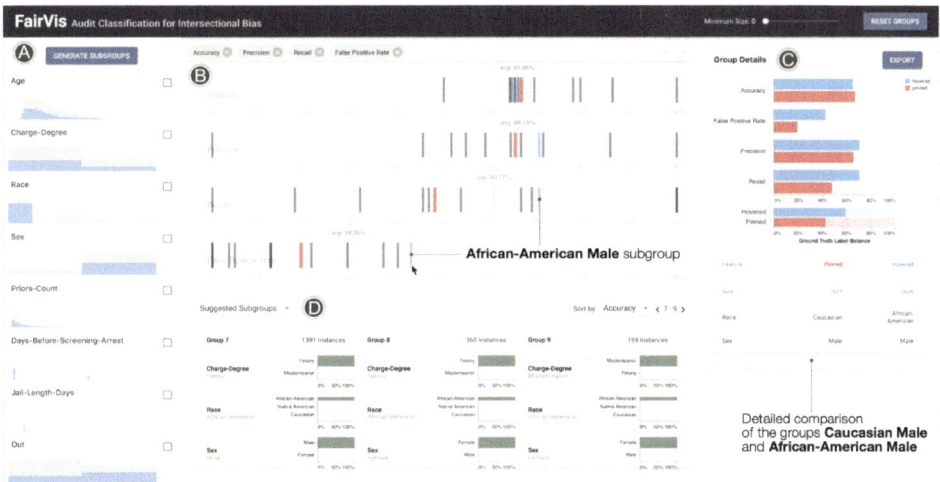

Fig. 5.10 The visualization of FairVis [26], a visual analytics tool that compares different data subgroups to identify potential bias in a developed model

interactively create sample subgroups based on a specific feature value or their combinations (Fig. 5.10a), while the tool can also suggest subgroups where the model may perform poorly or subgroups that are similar to the user-created ones (Fig. 5.10d). Besides, FairVis leverages strip plots to compare multiple subgroups on different fairness measures (Fig. 5.10b). It visualizes the distribution of each measure across all subgroups using a strip plot. In the strip plot, each subgroup is represented as a bar, with its horizontal position indicating the value of the measure. By comparing features and performance among the subgroups of interest, users can discover the reasons for the unfairness (Fig. 5.10c).

There are also other VIS4AI methods that utilize bar charts for similar purposes. Fair-Sight [4] and DiscriLens [245] use bar charts to represent feature distributions and juxtapose these bar charts for comparison. In addition, dot-based visualizations are also used to show feature distributions by replacing bars with dots for a more fine-grained analysis, where each dot represents a sample. Wexler et al. [254] developed the What-If tool that generates such dot-based visualizations for the analysis of fairness issues. For example, this tool customizes confusion matrices by placing dots in matrix cells to represent samples. The dots are colored according to the correctness of their predictions. These visualizations allow users to gain an overview of the prediction distributions and enable the identification of samples with unfair predictions. After selecting the samples of interest, users compare them with their neighboring samples to analyze the reason for the unfair predictions.

To identify the root cause for the different model behaviors in different sample subgroups, it is also important to understand the intersection and inclusion relationships between them. Despite the effectiveness of bar charts in comparing performance between subgroups, they cannot reveal such complex relationships. To tackle this problem, Wang et al. [246] designed an Euler-based visualization, RippleSet, to better reveal the complicated intersection and inclusion relationships across multiple subgroups in their work. As shown in Fig. 5.11, given a set of samples and their associated subgroups, the first step is to rearrange the samples into maximal inseparable sets. The samples in each inseparable set belong to the same set of subgroups. Each inseparable set is represented as a circle, and the samples in this set are represented as dots that are tightly packed and placed inside the corresponding circle. The circles that contain the samples belonging to the same subgroups are placed adjacently, which better represents the composition of subgroups while avoiding complex overlap between circles. For example, the sample subgroup D is composed of two adjacent circles representing the inseparable set CD and ABCD (upper part in Fig. 5.11d). This visualization helps users quickly understand the relationships between samples between different subgroups.

Another type of information that is useful for analyzing data-related fairness issues is the pairwise distances between samples. For example, FairSight [4] explains model fairness by examining how pairwise distances between samples are distorted from the input space to the output space. It uses a matrix to visualize the inter- and intra-subgroup distortion between samples from different data subgroups or in the same data subgroup. Each row and column represents a sample, and each cell represents the distortion between the corresponding

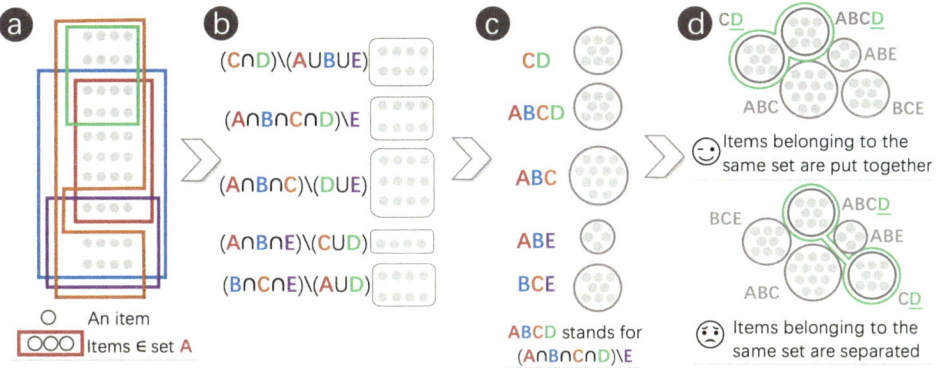

Fig. 5.11 The generation process of RippleSet [246]

samples. A lighter color in the cell indicates a smaller level of distortion, and vice versa. An unfair model often causes much distortion between samples in its prediction process, leading to a matrix consisting of many cells with dark colors.

The above visual analytics tools mainly focus on the fairness issue related to sample subgroups. In terms of the fairness issue related to individual samples, an effective analysis tool is the counterfactual explanation, which enables users to identify the causal features for model prediction. A counterfactual explanation for a specific sample involves finding a sample with a few modified features but a different prediction from the original one. Since such features result in a change in the model prediction, they reveal potential fairness issues. For example, if modifying the feature "gender" of a person will change the prediction of a resume screening model, the model is potentially biased toward gender.

Along this line, Cheng et al. [43] developed DECE (Fig. 5.12) to explore and understand the model behaviors by generating counterfactual explanations. It focuses on a specific kind of counterfactual sample, the r-counterfactual, which only changes one feature of the given sample. These samples enable users to study the effect of one specific feature at a time. To support the analysis of the r-counterfactuals, a table view and an instance view were developed. The table view, consisting of three parts, serves as an entry point for the analysis (Fig. 5.12a). The first part is a header that uses bar charts to show the distributions of the predicted classes and the features of the samples in the entire dataset (Fig. 5.12A1). In the first column, the colors encode the class labels, while the solid and striped bars encode the correct and wrong predictions, respectively. In each of the other columns, two histograms are used to display the feature distribution of the samples and their corresponding r-counterfactuals. The second part is a list of subgroups (Fig. 5.12A2). Each row uses the same visual encodings as the header, but visualizes the distributions of the corresponding subgroup instead of the entire dataset. The third part is an instance lens that shows the detailed information for the samples and the differences from their corresponding r-counterfactuals in a selected subgroup (Fig. 5.12A3). By investigating the subgroups and their corresponding

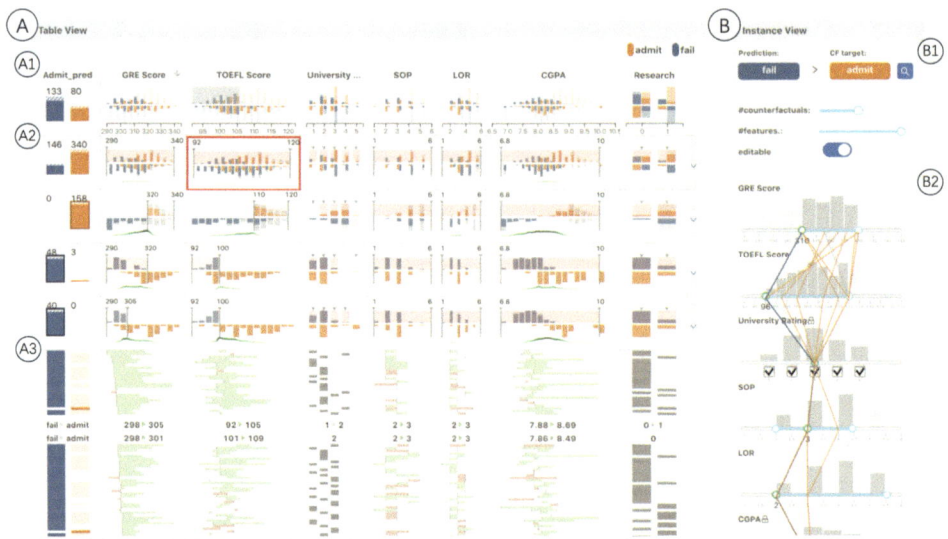

Fig. 5.12 The visualization of DECE [43], a visual analytics tool that enables the analysis of counterfactual explanations

r-counterfactuals, users form hypotheses and validate them in an instance view (Fig. 5.12b). The instance view utilizes an enhanced parallel horizontal axes, where each axis contains a histogram showing the distribution of the corresponding feature. For a selected sample, its r-counterfactuals are first calculated (Fig. 5.12B1). Each sample is then represented as a polyline, with the color encoding its predicted class, and the horizontal positions on the axes represent the feature values (Fig. 5.12B2).

In addition to DECE, several other VIS4AI methods also support counterfactual explanations [79, 254]. For example, Gomez et al. [79] developed ViCE to generate and visualize counterfactual explanations for binary classification tasks. They used a greedy algorithm to find counterfactual samples and used a set of columns to represent features and explain how their changes lead to the flip in the prediction.

Model-Related Fairness. Identifying fairness issues caused by the models is another important aspect. For classification models, such as random forest and logistic regression, an intuitive solution is to juxtapose several charts or plots to compare model performance, as adopted in the What-If Tool [254], DiscriLens [246], and FairVis [26]. For example, DiscriLens enables a side-by-side comparison of two models on a set of selected samples that potentially have fairness issues.

Recent VIS4AI work also studies the fairness of ranking models. For example, FairRankVis [262] explores the fairness issues in graph ranking models. To analyze the fairness of the ranking results between different models, FairRankVis designed three novel views. The first is the rank mapping view (Fig. 5.13a), which compares the changes in the ranking

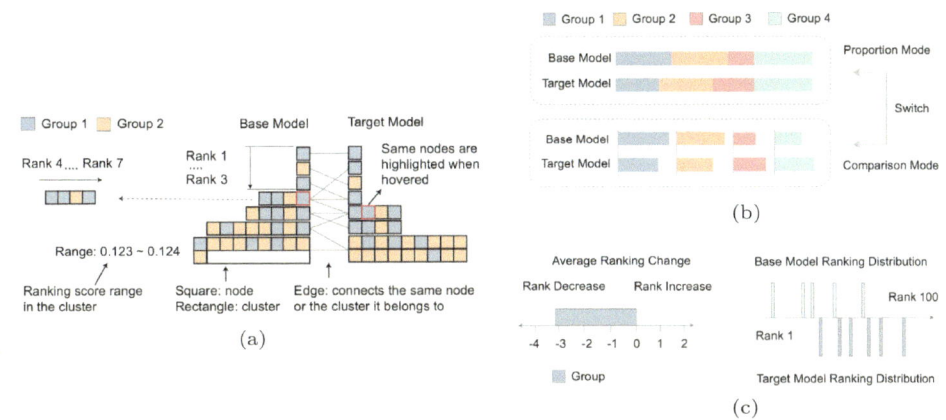

Fig. 5.13 FairRankVis [262], a visual analytics tool for comparing changes in ranking scores between different graph ranking models: **a** the rank mapping view, **b** the group proportion view, and **c** the group shift view

scores between different models for a set of samples. For each model, it first clusters these samples based on their ranking scores. Each sample is represented by a square, and those in the same cluster are horizontally stacked as a bar, where samples are placed from left to right in the descending order of their ranking scores. These bars are then vertically arranged based on the range of their ranking score. To better reveal the changes in the ranking of individual samples, the two sets of bars are placed side-by-side, and two bars are connected if there exists any sample that belongs to both bars. The second is the group proportion view (Fig. 5.13b). It features two sets of colored bars that represent the proportions of different groups in the top-ranking samples predicted by the two models. These bars can be stacked to show the distribution of group proportions in each model (proportion mode) or placed side-by-side to compare the group proportions between the models (comparison mode). The last is the group shift view (Fig. 5.13c). For each group, it includes a horizontal bar showing the average ranking change between the two models and a dual bar chart showing the detailed distributions of sample rankings in the two models, positioned above and below the x-axis, respectively.

5.3   Summary

This chapter categorizes the VIS4AI work in the model deployment stage into two parts based on their tasks: (1) decision explanation, and (2) model monitoring and maintenance. Accordingly, there are two main goals in the model deployment. The first goal is to enhance transparency and establish trust in model decisions. This is achieved through decision explanation techniques, which aim to provide clear and understandable visual explanations of the

Table 5.1 Summary of the VIS4AI studies in the model deployment stage

VIS4AI task	Goal	Papers
Decision explanation	Trustworthiness	[23, 46, 108, 173, 241, 279]
	Transparency	[125, 170, 171, 181, 216, 274, 275]
Model monitoring and maintenance	Robustness	[27, 50, 55, 142, 164, 185, 191, 249, 263]
	Fairness	[4, 26, 43, 78, 79, 245, 254, 262]

decision-making process. Simple visualizations, such as matrices, are preferred in this case, as they offer intuitive explanations that are easily understood by users with diverse backgrounds. The second goal is to ensure the robustness and fairness of the models through model monitoring and maintenance. To achieve this, model consumers often need to delve deeper into the models to diagnose the diverse issues. As a result, researchers tend to adopt and design customized visualizations to perform complicated analysis tasks. In addition, most visual analytics tools utilize multiple coordinated visualizations to support the analysis from multiple perspectives. For example, DriftVis [263] uses line charts to show temporal trends, scatterplots to display data distributions, and heat maps to identify when and where drifts occur. The multiple views work together to help identify the root causes of the drifts. We also observe that many issues in the model deployment stage involve the data and the models simultaneously. Thus, it would be meaningful to tightly integrate the analysis from both perspectives in the related VIS4AI work.

Many VIS4AI studies have been conducted to effectively achieve these tasks. For the convenience of the readers, we briefly summarize them based on the associated VIS4AI tasks in Table 5.1.

Research Challenges and Opportunities

6

VIS4AI presents many research challenges and opportunities across the entire machine learning lifecycle. At every stage, from data preparation to model development and model deployment, researchers and practitioners face unique hurdles that demand innovative visualization solutions to ensure efficient and effective AI systems. Beyond these stage-specific challenges, generic research challenges span the entire machine learning lifecycle. Furthermore, the prevalence of foundation models in recent years has added complexity to VIS4AI [268]. This demands even more innovative visualization methods to improve and adapt these models. In this chapter, we explore key challenges and potential avenues for further exploration.

6.1 Data Preparation

This section explores the challenges and opportunities associated with data preparation, including data quality research in weakly supervised learning and multi-modal learning, active sample selection, and explainable feature engineering.

6.1.1 Data Quality Research in Weakly Supervised Learning

Weakly supervised learning builds models from training data with quality issues such as inaccurate, insufficient, or inexact annotations [270, 282]. This presents the challenge of learning from imperfect annotations and highlights the necessity for new techniques to address such quality issues, with the ultimate goal of creating high-performance machine

---

Dr. Xiting Wang also contributes to this chapter.

learning models [144]. Most existing methods focus on **inaccurate** data quality issues (e.g., noisy annotations) and interactive labeling related to incomplete data quality issues (e.g., only some of the data are annotated). Less effort has been devoted to better exploitation of unannotated data or knowledge for insufficient data quality issues, as well as inexact data quality issues (e.g., coarse-grained labels). This opens avenues for potential future research on these topics.

First, the potential of visual analytics techniques to tackle the issue of **insufficient data** is not fully exploited. For example, improving the quality of unannotated data is critical for semi-supervised learning [143, 144]. The unannotated data can be combined with a small amount of annotated data during training to infer the correct dataset-to-annotation mapping. A typical example is graph-based semi-supervised classification, which depends on the similarity relationships between labeled and unlabeled data in the constructed kNN graph [143]. The effectiveness of this method relies on the quality of this kNN graph. However, automatic graph construction methods usually rely on global parameters (e.g., a global k value in the kNN graph construction method), which may not be locally suitable. Consequently, there is a need to employ visualization techniques to demonstrate how annotations propagate along the edges of the graph. This facilitates the understanding of how local graph structures impact model performance [37]. Based on this understanding, experts can adaptively modify the graph to gradually create a high-quality graph. Hence, an intriguing avenue for future research lies in exploring methods to extend the annotations from annotated data, particularly the detection boxes and segmentation masks, to unannotated data.

Second, the issue of **inexact data** quality, prevalent in real-world applications [282], has not received much attention from the visualization field. This issue refers to the situation where annotations are inexact, for example, coarse-grained annotations found in computed tomography scans. For these scans, annotations are typically derived from the corresponding diagnostic reports that confirm the presence of medical conditions such as tumors. However, they fail to pinpoint the specific slices or exact locations of these tumors. Although various machine learning methods [75, 281] have been proposed to learn from such coarse-grained annotations, the lack of exact information often leads to poor performance [144]. As a result, fine-grained validation is still required to improve data quality and model performance. Integrating interactive visualization with learning algorithms offers a promising avenue to address this issue. This integration facilitates an in-depth analysis of the root causes behind poor performance by examining the overall data distribution. Furthermore, it may support an interactive verification process designed to refine annotations and reduce expert efforts.

6.1.2 Data Quality Research in Multi-Modal Learning

Multi-modal learning uses multiple types of data, such as text, images, audio, video, and sensor data, to train machine learning models. This strategy improves performance by capturing the complementary information available in different data modalities. Consequently, the quality of multi-modal data is even more critical for building a high-performance model.

Enhancing the quality of multi-modal data faces challenges related to handling missing or noisy data, data fusion, and alignment of different modalities. VIS4AI is a promising technique to address these challenges. First, it is of fundamental importance to develop novel visual analytics techniques for assessing and improving the quality of multi-modal data, including reducing noise, handling missing values, and addressing inconsistencies. To achieve this, it is necessary to design metrics and algorithms that measure the quality of each modality and the overall data by considering key factors such as completeness, accuracy, consistency, and reliability. Second, it would be interesting to explore advanced visual analytics methods for effectively fusing information from multiple modalities, considering their inherent characteristics and the specific data quality issues of each modality; this involves developing visualization-assisted fusion strategies that maximize the benefits of combining modalities while addressing challenges such as heterogeneity, noise, and missing data. Finally, it is worth investigating visual analytics methods to align modalities and establish meaningful correspondences between different types of data, enabling effective analysis and modeling. This involves developing techniques to handle disparities in data distributions, formats, and semantics across modalities.

6.1.3   Active Selection of Training/Fine-Tuning/Test Data

The training and adaptation of large models are computationally intense and usually require millions or even billions of training data [187]. This large-scale data requirement introduces complexities in several aspects, including data storage, computational power, and processing time. Furthermore, the training of large models is becoming a serious source of carbon emissions that threaten our environment [218]. Recent studies have shown that selecting a subset of high-quality data for training can achieve comparable or even better performance [201, 276, 280]. This finding suggests the possibility of reducing the computational and environmental costs associated with model training. Visualizations serve as valuable tools for exploring large-scale datasets and selecting high-quality training data. However, there are two major challenges that need to be addressed.

The first challenge centers on scalability, a crucial issue for large models. The huge amount of data for training and fine-tuning these models often exceeds memory limits, which makes it difficult to process and visualize all the data at once. This situation not only necessitates out-of-memory sampling methods, but also poses real-time interaction challenges for visualization. One feasible solution is to present an overview of the entire dataset, which is created using a data subset obtained from these out-of-memory sampling methods. This strategy allows users to quickly examine the overall landscape and identify regions that warrant a closer examination. Focusing on these specific regions allows for a more detailed analysis. Given that the data is not stored in memory, exploring ways to facilitate real-time interactions becomes an essential research avenue.

The second challenge comes from the unannotated and unstructured nature of the training data, which hampers the evaluation of data quality and the selection of high-quality training samples. One potential solution involves designing multiple metrics to visually summarize the data characteristics from different perspectives. Furthermore, the unstructured characteristics pose difficulties for users in quickly understanding the content of samples, which requires innovative visualizations of the data to alleviate the cognitive load. Although multi-modal data is now widely used in training large models, the visualization of inter-modal alignments remains an under-researched area and deserves further investigation.

The selection of test data shares most of the challenges as the selection of training data such as scalability issues and the unstructured nature of the data. Despite these similarities, key differences exist and warrant attention. The primary goal of test data is to faithfully reflect the performance of models while also exposing their potential weaknesses. Therefore, it is crucial for test data to cover both common samples that models regularly process and "edge case" samples where models are prone to fail. Visualization techniques are suitable for examining the selection balance between these two types of samples. Therefore, it is worth exploring how to integrate visualization techniques with subset selection methods for a well-balanced selection.

6.1.4 Explainable Feature Engineering

Most existing studies for improving feature quality are designed for classical models and focus on tabular or textual data [155, 257]. The features of these data are naturally interpretable, which simplifies the process of feature engineering. The features extracted by deep models usually perform better than those that are manually crafted [57, 245]. However, the interpretability of deep features becomes challenging due to the black-box nature of deep models. This brings more obstacles to effective feature engineering.

First, the extracted features are obtained in a data-driven process, which may poorly represent the original images/videos when the datasets are biased. For example, given a dataset with only dark dogs and light cats, the extracted features may emphasize color and ignore other discriminating concepts such as shapes of faces and ears. Without a clear understanding of these biased features, it is hard to correct them comprehensively. Thus, an interesting topic for future work is to utilize interactive visualization to explore the root causes behind biased features. The key challenge here is how to measure the information preserved or discarded by the extracted features and present this information in an easily understandable visualization.

Second, redundancy exists in extracted deep features [10]. Removing redundant features offers several benefits such as reducing storage requirements and improving generalization [34]. However, without a clear understanding of the exact meaning of these features, it is hard to judge whether a feature is redundant. Therefore, an interesting future topic is to develop a VIS4AI method that intuitively communicates feature redundancy. This would enable experts to identify and then remove redundant features.

6.2   Model Development

This section explores the challenges and opportunities associated with model development, including the understanding of multi-modal models, model-agnostic explanations, online training diagnosis, interactive model refinement, and interactive model evaluation.

6.2.1   Understanding Multi-Modal Learning Models

Existing research on model analysis has achieved great success in understanding single-modal learning models. However, real-world applications often involve multi-modal data. For example, a physician diagnoses a patient by considering multiple types of data such as medical records (textual data), laboratory reports (tabular data), and computed tomography scans (images). When analyzing such multi-modal data, simply combining knowledge acquired from single-modal models falls short of capturing intricate relationships between different modalities. A more promising method is to use multi-modal machine learning techniques and leverage their capability to disclose insights across different data types. In this context, a powerful VIS4AI system is essential for understanding the output of these multi-modal learning models. Nowadays, many models have been developed to learn the joint representations of multi-modal data [12, 161]. An interesting future direction is how to effectively visualize the joint representations of multi-modal data in an all-in-one manner. This visualization can significantly enhance the understanding of the data and their relationships. Furthermore, various classical multi-modal tasks can be employed to enhance natural interactions in the visualization field. For example, in the vision-language scenario, the visual grounding task, which identifies the corresponding image area based on a textual description, can serve as a natural interface to facilitate natural-language-based image retrieval in a visual environment.

6.2.2   Model-Agnostic Explanations

As the machine learning field continues to evolve, there is a growing need for explanation techniques that can accommodate the increasing diversity and complexity of models. The traditional method of developing model-specific explanations for each individual model becomes challenging and time-consuming due to the sheer number and variety of models being developed. To address this challenge, model-agnostic explanations have emerged as a promising research direction in VIS4AI. Model-agnostic explanations aim to provide a unified and generic method that can be applied to a wide range of machine learning models, including classical models, deep models, and foundation models.

Developing model-agnostic explanation techniques presents a significant research opportunity for VIS4AI. Potential future research in VIS4AI involves visually revealing the underlying factors, patterns, and relationships that contribute to the model predictions without requiring in-depth knowledge of its internal workings. It also involves investigating new visualization techniques and interaction paradigms that allow users to explore and interact with the explanations effectively. Furthermore, exploring ways to integrate model-agnostic explanations into real-world applications and decision support systems can open avenues for enhancing human-machine collaboration and enabling more responsible and ethically sound AI systems.

6.2.3   Online Training Diagnosis

With the increasing complexity of models, the training time usually takes weeks or even months on high-end GPUs. Traditional offline methods gather relevant data after the training process and then feed them into the analysis tool, which is less effective in reducing unnecessary training trials. Moving visual analysis earlier in the model development workflow can save vast amounts of time and computing resources by halting ineffective and inefficient training immediately. Therefore, it is necessary to develop visualization techniques that are suitable for monitoring real-time running results and identifying performance issues and efficiency issues. Accordingly, there are two interesting avenues that deserve exploration.

The first promising avenue is to support an in-depth analysis of model performance during model training. While some existing efforts like Tensorboard [1] have supported the online monitoring of the training process, they only consider high-level performance metrics such as loss and prediction accuracy. These metrics are too abstract to effectively troubleshoot the reason why the model does not perform as expected. To address this issue, it is necessary to integrate advanced data analysis and model analysis modules into visualizations to provide richer information. By analyzing the sample content and how the model processes them, model developers can gain more insights into performance issues and address them accordingly.

The second promising avenue lies in the management of large-scale profiling data in online diagnosis. Given the rapid generation of profiling data and the input/output overhead associated with transferring data from GPU to memory or even disk storage, it becomes impractical to store all the data and then transfer them to the visualization tool for analysis. In situ visualization emerges as a promising method to address this issue [162], as it visualizes data directly within the computational environment where it is generated. Although in situ visualization has been shown to be useful in scientific visualization [202, 206], it is still underexplored whether it can be used to simplify and improve efficiency diagnosis in model training. Another possible solution is progressive visual analytics, which allows users to explore data incrementally [11, 226, 273]. Unlike traditional methods that handle entire datasets in one go, progressive visual analytics processes data incrementally, thereby boosting the efficiency of online analysis for large-scale profiling data produced incrementally in the training process. This offers a promising research avenue for assessing the effectiveness of online training.

6.2.4   Interactive Model Refinement

Recent studies have explored the use of prediction uncertainty to facilitate interactive model refinement [66, 147, 248, 264]. There are many methods to compute the prediction uncertainty such as relying on confidence scores generated by classifiers. In addition, visual hints can effectively guide users in examining samples with high prediction uncertainty. Following user refinement, the uncertainty scores will be recomputed. This allows users to iterate until they are satisfied with the prediction results. Such an interactive cycle opens avenues for improving the interactive model refinement process.

One promising direction is to interactively leverage the model results from previous iterations to guide refinement in subsequent ones. For example, in clustering applications, users can actively define must-link or cannot-link constraints between pairs of data samples based on the previous clustering results. These constraints improve model performance in the current iteration, including aligning the results more closely with the underlying data distribution and meeting the user-defined constraints. This user-driven constraint mechanism empowers the refinement process by allowing users to exert fine-grained control over the model behavior. Users can iteratively fine-tune the model based on their evolving insights and domain expertise to create a more customized and accurate representation of the data.

Another direction involves the integration of prior knowledge to identify areas that need refinement. This is particularly crucial for generative models, such as GANs and VAEs, where model-generated results may conflict with public or domain knowledge. Therefore, adopting a knowledge-based strategy can help identify inconsistencies between model-generated results and established knowledge. By detecting these inconsistencies, users can effectively refine these models by imposing constraints that align the prediction results with the established knowledge.

6.2.5   Interactive Performance Evaluation

Evaluating Free-Form Outputs. Recently, GPT series models have achieved impressive performance in various tasks, notably in answering open-ended questions without definitive ground-truth answers. However, evaluating the quality of free-form model responses remains challenging due to the high variability in possible responses and the absence of clear ground-truth answers. Addressing this challenge requires human involvement in the evaluation process. However, the sheer volume of data makes it unfeasible for users to manually inspect and assess each model response individually. One possible solution is to semi-automatically create rules for evaluating the model responses, which can be achieved by active learning methods. Visualizations enhance this process by offering a comprehensive overview of these evaluation rules and their associated model responses. Users can then iteratively refine these rules according to their preferences. This ultimately leads to more accurate and reliable evaluations. Another potential solution is to utilize visualizations to highlight the responses that are difficult for semi-automatic evaluation methods and present them to users for manual review. To minimize redundancy and simplify this process, it is essential to cluster a massive volume of responses and intuitively summarize the clustering results in visual forms.

Robustness of Deep Generative Models. Although recent deep generative models have demonstrated impressive generation abilities [25, 187], they can sometimes misinterpret input or generate off-target or even incorrect responses. Such inconsistencies pose challenges in the reliable deployment of these models, especially in scenarios where a single error could have significant consequences. As a result, there is an urgent need to gain a clear understanding of their robustness. With this understanding, users can assess how well these models may perform in different situations and identify weak areas that require fine-tuning to improve performance.

One possible solution to achieve this is to generate a set of input samples with perturbations and compare the corresponding model responses with well-designed visualizations. This method can well illustrate how small changes in the input can affect model responses. Thus, it provides insights into the robustness and sensitivity of the model. Visualizations provide an important way of identifying critical samples for closer examination, interactively constructing perturbed samples for deeper behavioral insights, and summarizing multiple model responses for efficient analysis. Another solution involves analyzing a large number of input samples collected in real-world scenarios to identify potential robustness issues. Models are often deployed in complex environments where they encounter a wide range of input samples. Manually examining each one for robustness issues can be an overwhelming task. Visualizations offer an efficient way to explore and filter a set of similar input samples that produce diverse output results. These anomalies often serve as indicators of potential robustness issues. Once these anomalous pairs are identified, visualization tools enable "what-if" analyses. These analyses examine how the model behaves under various conditions and then identify specific areas where the model robustness can be improved.

6.3   Model Deployment

This section explores the challenges and opportunities in model deployment, focusing on the evaluation of XAI explanations and the fitting of the dynamic nature of AI systems.

6.3.1 Evaluation of XAI Explanations

Evaluation of XAI explanations attracts much attention from researchers and practitioners. Current efforts focus on assessing factors such as explanation satisfaction, explanation quality, and scalability in XAI techniques [28, 97]. These efforts, mainly from the machine learning field, aim to improve user confidence and trust in machine learning models. However, they may not fully account for the user's cognitive processes and interpretability from a human-centered perspective. This gap highlights the need for robust evaluation metrics, which are fundamental ways to validate the effectiveness of XAI models. Without solid evaluation metrics and methods, any claim of model effectiveness lacks persuasive proof.

In the context of XAI, evaluation metrics enable a detailed measurement of how well a model meets the XAI requirements. Analogous to standard metrics such as accuracy, precision, and recall in classification, XAI metrics should validate model performance from specific XAI viewpoints. Recent discussions and efforts emphasize the importance of such metrics, which not only evaluate the performance of the XAI model but also improve the trust and confidence of the end user [97]. These metrics include qualitative and quantitative indicators that evaluate satisfaction with explanation, goodness, and scalability.

Despite the progress, there is an urgent need for more standardized and quantifiable XAI evaluation metrics to keep pace with the growing number of XAI tools and techniques. Introducing visualization techniques into the evaluation framework for XAI explanations becomes imperative. Such an integration promises a more comprehensive understanding of how users interact with and consume explanations, which potentially bridges the gap between technical validity and human intelligibility. Furthermore, novel visualization methods for qualitatively comparing different XAI methods should be considered in various application contexts.

6.3.2 Fitting the Dynamic Nature of AI Systems

The dynamic nature of AI systems refers to how machine learning models can learn, adapt, and evolve over time. As AI systems interact with new data, associated machine learning models can adjust and refine their functions to improve performance. This requires effective visualizations to appropriately represent these ongoing changes, which leads to the following challenges.

Concept Drift. Concept drift refers to changes in data distribution between the development and deployment stages, which often result in performance degradation. During the deployment stage, visualizations are essential as they can effectively illustrate when drifts occur and how they affect performance. Accordingly, an interesting future research area involves developing visualization techniques to effectively communicate drift and facilitate the root cause analysis of such occurrences. After analyzing concept drift, model developers also want to update the ML models in a timely manner to maintain high performance. Consequently, visualizations should offer real-time updates to reflect the evolving behavior of the models. This task becomes particularly challenging for models with frequent drifts.

Open-World Machine Learning. In addition to changes in data distribution, problem domains can also change in dynamic and evolving environments. For example, in the context of security checks, an object detector should not only accurately identify predefined objects, but also be able to detect unknown objects. Developing models that can handle both pre-defined and unknown classes is known as open-world machine learning. To improve model performance in open-world machine learning, it is necessary to quickly identify unknown classes of interest and then provide the necessary annotations. Visualizations serve as a useful tool for summarizing potential unknown classes and presenting them to model developers. After identifying the class of interest, VIS4AI techniques for data preparation can be used directly to provide valuable assistance in the annotation process.

Continual Learning. In real-world deployments, the accumulation of data over time and the evolution of tasks present unique challenges. Continual learning addresses these challenges by enabling models to dynamically adapt to these changes and update their knowledge while retaining previously learned knowledge. Replay-based methods, a straightforward and effective continual learning mechanism, preserve a representative subset of old data for reference during the acquisition of new knowledge. In this context, effective visualization of large datasets is essential. Techniques such as aggregation, summarization, and progressive rendering are crucial to visually analyzing large, dynamic data. These methods not only provide deep insights into dynamic data, but also help identify critical data that should be preserved for future replay. Another type of method involves the use of regularization-based techniques. These techniques integrate regularization terms into the learning process to achieve a balance between preserving old knowledge and learning new knowledge. Visualization also plays a key role here by facilitating the assessment and analysis of the model performance across the old and new tasks. The insights acquired by interactive exploration guide the model toward a more effective balance between various tasks and ensure its robust adaptation to evolving data and tasks.

6.4   Generic Challenges and Opportunities

This section explores the generic research challenges that span the entire machine learning lifecycle, including building trust from XAI explanations and cross-cultural and ethical considerations.

6.4.1   Building Trust from XAI Explanations

According to Hoffman et al. [97], clear and comprehensive XAI explanations about a learning model foster appropriate trust. These explanations allow users to understand the model behavior more accurately and lead to better user performance and model performance. The endeavor to establish this trust should not be centered on achieving a single, unchanging condition or maintaining a numerical value that lacks contextual relevance. Instead, it should aim to foster an appropriate expectation that is dependent on the context. As a result, the process of building trust based on explanations is inherently an exploratory endeavor, which is well aligned with the nature of interactive visualization. However, leveraging visualization to build trust is not straightforward due to various contexts and potential biases.

The first challenge involves maintaining contextual relevance. The goal is to ensure that the used visualizations are relevant to the context of the AI application. For example, if an AI system is used for medical diagnosis, visualizations should ideally be in a format familiar to healthcare professionals such as charts, graphs, and anatomical drawings. They should also represent information relevant to medical decision-making such as symptom severity or probability of different diagnoses. Such contextual relevance is critical because if the visualizations are not suited to the user's context, including their needs, background, and the specific problem they are trying to solve, they may be confusing, unhelpful, or even misleading. Therefore, when designing visualizations, it is crucial to consider the intended users and their context to tailor the visualizations accordingly. As more and more AI systems need to enhance trust in different applications, it is worth studying automatic methods to generate visualizations that can adapt to different contexts. Moreover, these visualizations need to accurately represent the inner workings of the AI system because any misrepresentation could lead to misplaced trust.

The second challenge relates to the bias introduced by visualizations, which poses a significant barrier to fostering trust in XAI. Visualizations might unintentionally bias toward or against certain results or interpretations, potentially due to design bias, data bias, and interpretation bias. Design bias can occur when the design of the visualization unintentionally leads users to interpret the data in a certain way. For example, the choice of colors, scale, or the emphasis on certain data points over others can cause misunderstanding of the information presented. If the underlying data used by the AI is biased, the visualization will likely reflect and possibly amplify these biases. For example, if the data used to train the AI system has a racial, gender, or socio-economic bias, this bias could be visually presented, leading to skewed understanding and mistrust. Additionally, interpretation bias can occur when users bring their own preconceived notions, beliefs, or biases to their interpretation of the visualization. For example, a user might interpret a visualization in a way that confirms their existing beliefs (confirmation bias), or they might focus more on information that is visually prominent, even if it is not the most important (salience bias). To overcome these biases, it is important to adopt ethical and transparent practices in both data handling and visualization design. This includes being aware of potential biases, testing visualizations with various user groups, and being open to feedback and adjustments. Another potential solution is to apply bias mitigation techniques in AI such as fair representation learning [48], adversarial debiasing [101], and calibration [198]. These techniques can assist in minimizing the bias present in the decisions made by the AI system, which are subsequently displayed through visualizations.

6.4.2   Cross-Cultural and Ethical Considerations

As AI systems are being increasingly deployed in various cultural contexts and used by different user groups, it is crucial to prioritize culturally sensitive, ethically sound, and socially aligned explanations through VIS4AI techniques. Consequently, it is necessary to explore how VIS4AI techniques can effectively navigate cross-cultural differences, address ethical dilemmas, and assess their broader societal impact. Such research is key to advancing the area of VIS4AI and promoting inclusive and responsible AI deployments.

First, cross-cultural differences can significantly impact how people perceive and interpret information. Cultural factors, such as languages, beliefs, values, and norms, influence the understanding and acceptance of AI systems and their explanations. Therefore, it is important to investigate how VIS4AI techniques can account for and adapt to cross-cultural differences in explanation generation and presentation. This involves studying cultural biases in AI systems, developing culturally aware explanation methods, and conducting user studies in various cultural contexts to assess the effectiveness and appropriateness of VIS4AI techniques.

Second, ethics play a crucial role in the development and use of AI systems. VIS4AI techniques should align with ethical principles such as fairness, transparency, accountability, and privacy. This involves addressing issues such as algorithmic bias, discrimination, and the potential impact of VIS4AI explanations on vulnerable populations. Investigating ethical frameworks and guidelines for VIS4AI is essential to ensure the responsible and ethical deployment of AI systems with visual explanations.

Third, addressing cross-cultural and ethical considerations with VIS4AI requires interdisciplinary collaboration and engagement with experts in areas such as cultural studies, ethics, and social sciences. It involves incorporating diverse perspectives, conducting user-centered research, and considering the broader social impact of VIS4AI techniques. These efforts lead to research possibilities for developing inclusive frameworks and methods that integrate these varied insights. This integration promotes advances in VIS4AI that are culturally aware and ethically grounded.

6.4.3   Dynamic Explanations

Static explanations may not fully meet user needs or understand complex AI systems. This presents a need for more dynamic, real-time explanatory frameworks. Research into dynamic explanation methods that facilitate real-time interaction with the ML model can significantly enrich the interpretability and user experience. However, crafting dynamic explanations that effectively accommodate the evolving nature of AI decision-making poses its own set of challenges. It demands a sophisticated balance between the capacity to reflect real-time changes, clarity, and accuracy.

One promising research direction is the exploration of real-time visualization techniques that provide continuous updates on the decision-making process of models as new data is processed. This involves developing visualizations that dynamically highlight the most influential features or decision rules in response to user queries or changes in the input data. It may also involve providing real-time visual feedback, helping users understand the impact of their interactions or modifications on the model output. Consequently, users gain a deeper understanding of the system's behavior.

Another area of future research is the development of interactive explanation interfaces that allow users to manipulate and explore different aspects of the ML model. By enabling users to actively interact with the model, they can gain insights into its decision boundaries and sensitivity to different factors, and explore "what-if" scenarios to understand the system's behavior in different contexts.

Furthermore, investigating techniques to incorporate user feedback into the explanation process can be a promising avenue for future research. By allowing users to provide feedback on explanations, such as rating their comprehensibility or relevance, the system can learn and adapt its explanation strategies to better align with user needs. This iterative feedback loop between users and the AI system can progressively improve the quality and relevance of the explanations over time.

6.5   Foundation Models

Recent research has shown that foundation models, which are adaptable to a wide range of downstream tasks, have become a general paradigm for building AI systems [22]. They typically possess parameters ranging from nearly a billion to even trillions and are trained on extensive datasets. Such versatility positions them as a general paradigm for building AI systems, thereby deeply influencing various aspects of human life [22]. According to a recent OpenAI report, around 19% of jobs have witnessed a significant transformation, and at least half of their tasks are affected by these models [68]. Although these models have shown impressive performance in various AI tasks, the sheer scale and complexity of recent iterations introduce new obstacles to meeting different human needs in different downstream tasks. Overcoming these challenges requires innovative human-in-the-loop methods to maximize their potential. VIS4AI techniques align perfectly with these needs and have become indispensable in the pre-training and adaptation of these foundation models, with a particular focus on improving data quality and developing models. The integration between VIS4AI techniques and foundation models is illustrated in Fig. 6.1. As we have already discussed

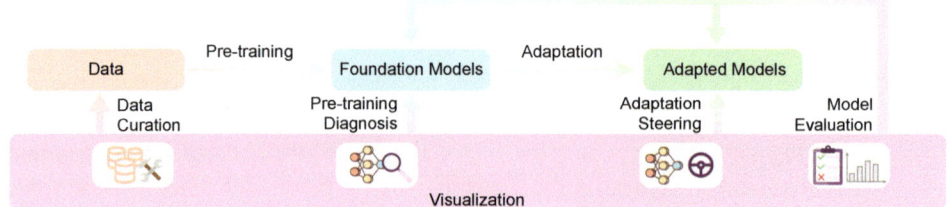

Fig. 6.1 How visualizations enhance foundation models along the learning pipeline

data curation in Sect. 6.1, this section concentrates on research challenges and opportunities related to foundation model development, including pre-training diagnosis and adaptation steering.

6.5.1   Pre-Training Diagnosis

The intrinsic nature of foundation models is defined by their vast number of parameters. While this vastness is the source of their capability, it also makes them difficult to interpret. Understanding the complex interactions, transformations, and computations within these numerous parameters presents a challenge. When a foundation model produces outputs, it is the result of a cascade of intricate operations influenced by millions or even billions of parameters. Tracing back through these operations to identify the exact reasoning or mechanism is similar to navigating a vast, complex maze without a map. As a model grows in **complexity and size**, understanding the specific factors or processes that contribute to its output becomes increasingly difficult.

The aforementioned challenge posed by the complexity and scale of foundation models demands innovative visualization solutions to incorporate human knowledge into the analysis process. These visualization tools can serve as "lenses" that allow users to look into the intricacies of these models and offer insights that can be grasped intuitively. Furthermore, employing rich interactions for exploration is also important in explaining foundation models. These exploration methods aim to distill the complex behaviors of foundation models into more understandable forms without compromising their essence. This might involve developing multi-level interpretation mechanisms, where users can select the granularity of the explanation, harness unsupervised techniques to automatically identify the most salient features or operations driving the model decisions, and present them for further analysis.

Multi-level interpretation mechanisms are tailored to offer explanations at varying levels of detail, from a high-level overview to detailed, granular insights. At the highest level, these explanations provide a general summary of the decision-making logic of the model. This is referred to as *surface-level interpretation*. For example, for a text generation task, a surface-level explanation could state, "The model generated this sentence based on the overall sentiment of the input." It can also provide a summary of the associated statistics such as confidence and bias scores. The next level can provide *component-level interpretation*, which aims to explain the role of specific components of the model such as particular layers or attention heads. For example, "The 10th attention head focused primarily on the relationship between subject and object in the sentence." The potential deepest level can provide *parameter-level interpretation*. It enables the examination of the influence and interactions of specific parameters or groups of parameters. This involves visualizing weights, gradients, or activations associated with particular tokens or features.

Given the vast amount of data present at each level, there is also a pressing need for an effective sampling method that can easily capture human interest and represent the corresponding data. This motivates the study of interactive sampling strategies, which requires the development of interactive visualizations to facilitate the detection of different user intents and then provide tailored subsets of data. These strategies enable users to navigate through complex data layers seamlessly. For example, they probe deeper into specific areas of interest or step back for a broader view, which enhances the overall understanding of the model functioning.

6.5.2 Adaptation Steering

Model Fine-tuning. After fine-tuning a foundation model for a specific task, the model will deviate from its pre-trained version. The changes include not only performance metrics, but also variations in model behavior such as processing different types of input and developing new input-output associations. By analyzing these behavior changes, model developers can understand how generic knowledge evolves into task-specific knowledge and identify where the model does not work as expected. Therefore, a promising research opportunity lies in using visualizations to effectively monitor these behavior changes and reveal abnormal behavior during the fine-tuning process. With a deep understanding of the behavior changes, model developers can identify when the model starts to exhibit biases or vulnerabilities that downgrade model performance. After identifying these negative issues, visualizations offer an efficient way to interactively steer the fine-tuning process, for example, by adding more balanced or targeted data. This method improves not only model performance but also reliability and robustness.

Prompt Engineering. Recent studies have shown that providing some high-quality examples within the prompts can greatly enhance model performance, which is known as the in-context learning ability [58]. In-context learning is a valuable component of prompt engineering. In this setup, prompt engineering becomes a critical exercise in curating and structuring examples that can guide the model effectively. To fully leverage the capabilities of foundation models and achieve satisfactory performance, the provided examples should be well suited for the downstream task. However, selecting high-quality examples requires expertise and often involves iterative refinement, which is usually trial-and-error in nature.

Visualizations offer an efficient way to facilitate this refinement process by integrating humans into the analysis loop [44, 156, 270]. One promising solution involves employing visualizations to visually illustrate model responses across different in-context examples. The insights derived from the visualizations enable users to evaluate the effectiveness of the examples and identify those most effective for the current task. Based on the findings, they can make informed refinements to the examples for better performance. In addition to interactively refining examples for each task, another promising direction lies in using visualizations to summarize general principles for example selection. By exploring different subsets of examples and conducting a comparative analysis among them, users can summarize the principles for determining which types of examples are beneficial and which are not. These principles contribute to a more systematic and informed example selection to craft effective prompts for the downstream task.

Alignment via Human Feedback. In the model adaptation process, aligning the model behavior with human preferences is essential. This alignment not only improves the user experience by generating more relevant responses, but also addresses ethical and social concerns [188]. Recently, reinforcement learning from human feedback has been shown to be effective in aligning model behavior with human preferences [188]. This method first trains a reward model directly from human feedback, which predicts whether the response aligns with human preferences (high reward) or not (low reward). Subsequently, this reward information guides the optimization of the foundation models through reinforcement learning. This process faces two primary challenges. First, collecting high-quality human feedback is a difficult task in itself, and the difficulty is amplified when the data must be fed into a reward model that drives reinforcement learning. Any errors or biases in feedback collection can result in skewed training and unreliable models. Second, the reward models might not be correctly updated to reflect human preference after receiving feedback. Ensuring the alignment between reward models and actual human preferences is still a challenging task. Visualization techniques are suitable for both tasks. First, interactive visualizations have already demonstrated their value in enhancing the process of collecting human feedback and ensuring their quality. For example, existing research on interactive data labeling shows the effectiveness of visualization techniques in facilitating the collection of human-generated data [17, 120, 265]. Second, visualizations offer an efficient way to diagnose the training process of reward models and interactively refine them through additional human feedback. The tight integration of human feedback into this process better aligns reward models with actual human preferences. This integration leads to more accurate and reliable reward information for the ongoing optimization of the foundation model.

Model Exploration and Selection. Recently, there has been an increasing trend among users to upload their foundation models together with meta-data (e.g., descriptions, model architectures, resources requirements) to learnware markets [2, 105, 283]. The surge in the availability of publicly accessible foundation models paves the way for more efficient AI system development. When confronted with an AI task, users have the option of searching and selecting a pre-existing model in the learnware market that fits their specific needs.

However, without sufficient expertise, navigating the expansive model space to find the most suitable foundation model can be challenging to them. The key challenge lies in facilitating user exploration by capturing user needs and recommending high-performance models. One potential solution is to employ visualization techniques to visually illustrate the model space. Through these visualizations, users can navigate the complex model space more easily, understand model behaviors, identify model limitations, and compare models from multiple perspectives such as performance metrics and resource requirements. Such a comprehensive understanding and comparison enable them to identify the most suitable model for their specific tasks.

Conclusion

As AI continues to play a vital role in high-stakes domains such as precision medicine, law enforcement, and financial investment, the generalization, interpretability, and trustworthiness of machine learning models have become paramount. VIS4AI emerges as a powerful method that combines machine learning techniques with interactive visualization to address these challenges. By covering every stage of the machine learning process, VIS4AI techniques focus on improving data quality and feature quality, as well as improving model development and deployment. Through interactive visualizations, users can gain deeper insights into the underlying data, models, and predictions, enabling better decision-making.

The book provides a comprehensive overview of the conceptual framework of VIS4AI. It covers a wide range of visualization techniques that can be applied for different purposes and scenarios. In addition, it explores the challenges and opportunities in the VIS4AI area, and highlights topics for further research and innovation. The proposed research topics range from improving the quality of training data to developing unified methods for evaluating models using interactive visualization and interactively pre-training and adapting foundation models. These topics of investigation have the potential to advance the field and improve the performance and trustworthiness of AI systems.

With its extensive coverage and practical insights on VIS4AI, this book is a useful resource for researchers and practitioners. It helps readers understand how visualization and artificial intelligence work together to solve real-world problems. By reading this book, readers will gain a deeper understanding of the intersection between visualization and artificial intelligence. They are also equipped with the knowledge and tools they need to address the challenges and take advantage of the opportunities that arise.

S. Liu et al., *Visualization for Artificial Intelligence*, Synthesis Lectures on Visualization, https://doi.org/10.1007/978-3-031-75340-4

Appendix

Acronyms

kNN	k-nearest neighbors
AE	Autoencoders
AI	Artificial Intelligence
BERT	Bidirectional Encoder Representations from Transformer
CLIP	Contrastive Language-Image Pertaining
CNN	Convolutional Neural Network
DAG	Directed Acyclic Graph
DGM	Deep Generative Model
DNN	Deep Neural Network
DQN	Deep Q-learning Network
DRL	Deep Reinforcement Learning
GAN	Generative Adversarial Network
GPT	Generative Pre-trained Transformer
GSSL	Graph-based Semi-Supervised Learning
LSTM	Long Short-Term Memory
ML	Machine Learning
MLP	Multi-Layer Perceptron
NLP	Natural Language Processing
OoD	Out-of-Distribution
PCA	Principal Component Analysis
PCP	Parallel Coordinate Plots
POS	Part-Of-Speech
RNN	Recurrent Neural Networks
SCAN	Symbol Concept Association Networks

© The Editor(s) (if applicable) and The Author(s), under exclusive license to Springer   127
Nature Switzerland AG 2025
S. Liu et al., *Visualization for Artificial Intelligence*, Synthesis Lectures on Visualization,
https://doi.org/10.1007/978-3-031-75340-4

SOM   Self-Organizing Map
SVM   Support Vector Machine
t-SNE   t-distributed Stochastic Neighbor Embedding
VAE   Variational Autoencoders
VIS$_4$AI Visualization for Artificial Intelligence
ViT   Vision Transformer
XAI   Explainable Artificial Intelligence.

References

1. Abadi M, Agarwal A, Barham P, et al (2016) Tensorflow: Large-scale machine learning on heterogeneous distributed systems. CoRR abs/1603.04467. https://doi.org/10.48550/arXiv.1603.04467
2. AdapterHub (2024) Adapterhub. https://adapterhub.ml/, Last accessed 2024-01-01
3. Afzal S, Chaudhary A, Gupta N, et al (2021) Data-debugging through interactive visual explanations. In: Trends and Applications in Knowledge Discovery and Data Mining: PAKDD 2021 Workshops, WSPA, MLMEIN, SDPRA, DARAI, and AI4EPT, Delhi, India, May 11, 2021 Proceedings 25, Springer, pp 133–142
4. Ahn Y, Lin YR (2020) FairSight: Visual analytics for fairness in decision making. IEEE Transactions on Visualization and Computer Graphics 26(1):1086–109. https://doi.org/10.1109/tvcg.2019.2934262
5. Alemzadeh S, Niemann U, Ittermann T, et al (2020) Visual analysis of missing values in longitudinal cohort study data. Computer Graphics Forum 39(1):63–75
6. Alsallakh B, Hanbury A, Hauser H, et al (2014) Visual methods for analyzing probabilistic classification data. IEEE Transactions on Visualization and Computer Graphics 20(12):1703–171. https://doi.org/10.1109/tvcg.2014.2346660
7. Arbesser C, Spechtenhauser F, Mühlbacher T, et al (2017) Visplause: Visual data quality assessment of many time series using plausibility checks. IEEE Transactions on Visualization and Computer Graphics 23(1):641–650
8. Arrieta AB, Díaz-Rodríguez N, Del Ser J, et al (2020) Explainable artificial intelligence (XAI): Concepts, taxonomies, opportunities and challenges toward responsible AI. Information fusion 58:82–115
9. Artur E, Minghim R (2019) A novel visual approach for enhanced attribute analysis and selection. Computers & Graphics 84:160–172
10. Ayinde BO, Zurada JM (2018) Building efficient convnets using redundant feature pruning. In: Proceedings of the International Conference on Learning Representations

S. Liu et al., *Visualization for Artificial Intelligence*, Synthesis Lectures on Visualization,
https://doi.org/10.1007/978-3-031-75340-4

11. Badam SK, Elmqvist N, Fekete JD (2017) Steering the craft: Ui elements and visualizations for supporting progressive visual analytics. In: Computer Graphics Forum, Wiley Online Library, pp 491–502

12. Baltrušaitis T, Ahuja C, Morency LP (2018) Multimodal machine learning: A survey and taxonomy. IEEE Transactions on Pattern Analysis and Machine Intelligence 41(2):423–443

13. Bäuerle A, Neumann H, Ropinski T (2020) Classifier-guided visual correction of noisy labels for image classification tasks. Computer Graphics Forum 39(3):195-205

14. Bäuerle A, van Onzenoodt C, Ropinski T (2021) Net2Vis–a visual grammar for automatically generating publication-tailored cnn architecture visualizations. IEEE Transactions on Visualization and Computer Graphics 27(6):2980–2991

15. Bäuerle A, van Onzenoodt C, der Kinderen S, et al (2022) Where did my lines go? visualizing missing data in parallel coordinates. In: Computer Graphics Forum, Wiley Online Library, pp 235–246

16. Berger M (2020) Visually analyzing contextualized embeddings. In: 2020 IEEE Visualization Conference (VIS), IEEE, pp 276–280

17. Bernard J, Zeppelzauer M, Lehmann M, et al (2018) Towards user-centered active learning algorithms. Computer Graphics Forum 37(3):121–132

18. Bertucci D, Hamid MM, Anand Y, et al (2023) Dendromap: Visual exploration of large-scale image datasets for machine learning with treemaps. IEEE Transactions on Visualization and Computer Graphics 29(1):320–330. https://doi.org/10.1109/TVCG.2022.3209425

19. Bilal A, Jourabloo A, Ye M, et al (2018) Do convolutional neural networks learn class hierarchy? IEEE Transactions on Visualization and Computer Graphics 24(1):152–16. https://doi.org/10.1109/tvcg.2017.2744683

20. Bishop CM, Nasrabadi NM (2006) Pattern recognition and machine learning. Springer

21. Bogl M, Aigner W, Filzmoser P, et al (2013) Visual analytics for model selection in time series analysis. IEEE Transactions on Visualization and Computer Graphics 19(12):2237–224. https://doi.org/10.1109/tvcg.2013.222

22. Bommasani R, Hudson DA, Adeli E, et al (2021) On the opportunities and risks of foundation models. arXiv preprint arXiv:2108.07258

23. Broeksema B, Baudel T, Telea A, et al (2013) Decision exploration lab: A visual analytics solution for decision management. IEEE Transactions on Visualization and Computer Graphics 19(12):1972–1981. https://doi.org/10.1109/tvcg.2013.146

24. Brooks M, Amershi S, Lee B, et al (2015) FeatureInsight: Visual support for error-driven feature ideation in text classification. In: Proceedings of the IEEE Conference on Visual Analytics Science and Technology, pp 105–112

25. Brown TB, Mann B, Ryder N, et al (2020) Language models are few-shot learners. 2005.14165

26. Cabrera ÁA, Epperson W, Hohman F, et al (2019) Fairvis: Visual analytics for discovering intersectional bias in machine learning. In: 2019 IEEE Conference on Visual Analytics Science and Technology (VAST), IEEE, pp 46–56

27. Cao K, Liu M, Su H, et al (2021) Analyzing the noise robustness of deep neural networks. IEEE Transactions on Visualization and Computer Graphics 27(7):3289–3304

28. Carvalho DV, Pereira EM, Cardoso JS (2019) Machine learning interpretability: A survey on methods and metrics. Electronics 8(8):832

29. Cashman D, Patterson G, Mosca A, et al (2018) RNNbow: Visualizing learning via backpropagation gradients in RNNs. IEEE Computer Graphics and Applications 38(6):39–5. https://doi.org/10.1109/mcg.2018.2878902

30. Cashman D, Perer A, Chang R, et al (2020) Ablate, variate, and contemplate: Visual analytics for discovering neural architectures. IEEE Transactions on Visualization and Computer Graphics 26(1):863–873. https://doi.org/10.1109/tvcg.2019.2934261

31. Cashman D, Xu S, Das S, et al (2021) CAVA: A visual analytics system for exploratory columnar data augmentation using knowledge graphs. IEEE Transactions on Visualization and Computer Graphics 27(2):1731–1741

32. Cavallo M, Demiralp Ç (2018) Track xplorer: A system for visual analysis of sensor-based motor activity predictions. Computer Graphics Forum 37(3):339–34. https://doi.org/10.1111/cgf.13424

33. Cavallo M, Demiralp Ç (2019) Clustrophile 2: Guided visual clustering analysis. IEEE Transactions on Visualization and Computer Graphics 25(1):267–276

34. Chandrashekar G, Sahin F (2014) A survey on feature selection methods. Computers & Electrical Engineering 40(1):16–28

35. Chatzimparmpas A, Martins RM, Kucher K, et al (2022) Featureenvi: Visual analytics for feature engineering using stepwise selection and semi-automatic extraction approaches. IEEE Transactions on Visualization and Computer Graphics 28(04):1773–1791. https://doi.org/10.1109/TVCG.2022.3141040

36. Chen C, Jiang L, Lei N, et al (2020) An interactive feature selection method based on learning-from-crowds. SCIENTIA SINICA Informationis 50(6):794–812. https://doi.org/10.1360/SSI-2019-0208, URL http://www.sciengine.com/publisher/ScienceChinaPress/journal/SCIENTIASINICAInformationis/50/6/10.1360/SSI-2019-0208

37. Chen C, Wang Z, Wu J, et al (2021a) Interactive graph construction for graph-based semi-supervised learning. IEEE Transactions on Visualization and Computer Graphics 27(9):3701–3716

38. Chen C, Yuan J, Lu Y, et al (2021b) OoDAnalyzer: Interactive analysis of out-of-distribution samples. IEEE Transactions on Visualization and Computer Graphi. https://doi.org/10.1109/TVCG.2020.2973258

39. Chen C, Wu J, Wang X, et al (2022) Towards better caption supervision for object detection. IEEE Transactions on Visualization and Computer Graphics 28(4):1941–1954

40. Chen C, Chen J, Yang W, et al (2024a) Enhancing single-frame supervision for better temporal action localization. IEEE Transactions on Visualization and Computer Graphics To be published

41. Chen C, Guo Y, Tian F, et al (2024b) A unified interactive model evaluation for classification, object detection, and instance segmentation in computer vision. IEEE Transactions on Visualization and Computer Graphics 30(1):76–8. https://doi.org/10.1109/TVCG.2023.3326588

42. Chen S, Yuan X, Wang Z, et al (2016) Interactive visual discovering of movement patterns from sparsely sampled geo-tagged social media data. IEEE Transactions on Visualization and Computer Graphics 22(1):270–279

43. Cheng F, Ming Y, Qu H (2021) DECE: Decision explorer with counterfactual explanations for machine learning models. IEEE Transactions on Visualization and Computer Graphics 27(2):1438–1447

44. Choo J, Liu S (2018) Visual analytics for explainable deep learning. IEEE Computer Graphics and Applications 38(4):84–92

45. Choo J, Lee C, Reddy CK, et al (2013) UTOPIAN: User-driven topic modeling based on interactive nonnegative matrix factorization. IEEE Transactions on Visualization and Computer Graphics 19(12):1992–2001. https://doi.org/10.1109/tvcg.2013.212

46. Collaris D, van Wijk JJ (2020) ExplainExplore: Visual exploration of machine learning explanations. In: Proceedings of the IEEE Pacific Visualization Symposium, pp 26–3. https://doi.org/10.1109/pacificvis48177.2020.7090

47. Cortes C, Jackel LD, Chiang WP (1995) Limits on learning machine accuracy imposed by data quality. In: Advances in Neural Information Processing Systems, pp 239–246

48. Creager E, Madras D, Jacobsen JH, et al (2019) Flexibly fair representation learning by disentanglement. In: International conference on machine learning, pp 1436–1445

49. Cui W, Liu S, Tan L, et al (2011) TextFlow: Towards better understanding of evolving topics in text. IEEE Transactions on Visualization and Computer Graphics 17(12):2412–2421

50. Das N, Park H, Wang ZJ, et al (2020) Bluff: Interactively deciphering adversarial attacks on deep neural networks. In: 2020 IEEE Visualization Conference (VIS), pp 271–275

51. Das S, Cashman D, Chang R, et al (2019) BEAMES: Interactive multimodel steering, selection, and inspection for regression tasks. IEEE Computer Graphics and Applications 39(5):20–3. https://doi.org/10.1109/mcg.2019.2922592

52. Deng L (2012) The MNIST database of handwritten digit images for machine learning research. IEEE Signal Processing Magazine 29(6):141–142

53. DeRose JF, Wang J, Berger M (2021) Attention flows: Analyzing and comparing attention mechanisms in language models. IEEE Transactions on Visualization and Computer Graphics 27(2):1160–1170

54. Devlin J, Chang MW, Lee K, et al (2019) BERT: Pre-training of deep bidirectional transformers for language understanding. In: Proceedings of the 2019 Conference of the North American Chapter of the Association for Computational Linguistics: Human Language Technologies, Volume 1 (Long and Short Papers). Association for Computational Linguistics, Minneapolis, Minnesota, pp 4171–418. https://doi.org/10.18653/v1/N19-1423, URL https://aclanthology.org/N19-1423

55. Diehl A, Pelorosso L, Delrieux C, et al (2017) Albero: A visual analytics approach for probabilistic weather forecasting. Computer Graphics Forum 36(7):135–14. https://doi.org/10.1111/cgf.13279

56. Dingen D, vant Veer M, Houthuizen P, et al (2019) RegressionExplorer: Interactive exploration of logistic regression models with subgroup analysis. IEEE Transactions on Visualization and Computer Graphics 25(1):246–255. https://doi.org/10.1109/tvcg.2018.2865043

57. Donahue J, Jia Y, Vinyals O, et al (2014) DeCAF: A deep convolutional activation feature for generic visual recognition. In: Proceedings of the International Conference on Machine Learning, pp 647–655

58. Dong Q, Li L, Dai D, et al (2022) A survey for in-context learning. arXiv preprint arXiv:2301.00234

59. Dorigo M, Gambardella LM (1997) Ant colony system: a cooperative learning approach to the traveling salesman problem. IEEE Transactions on evolutionary computation 1(1):53–66

60. Dosovitskiy A, Beyer L, Kolesnikov A, et al (2020) An image is worth 16x16 words: Transformers for image recognition at scale. arXiv preprint arXiv:2010.11929

61. Dou W, Yu L, Wang X, et al (2013) HierarchicalTopics: Visually exploring large text collections using topic hierarchies. IEEE Transactions on Visualization and Computer Graphics 19(12):2002–2011. https://doi.org/10.1109/tvcg.2013.162

62. Dowling M, Wenskovitch J, Fry J, et al (2019) Sirius: Dual, symmetric, interactive dimension reductions. IEEE Transactions on Visualization and Computer Graphics 25(1):172–182

63. Eichner C, Schumann H, Tominski C (2019) Making parameter dependencies of time-series segmentation visually understandable. Computer Graphics Forum 39(1):607–62. https://doi.org/10.1111/cgf.13894

64. El-Assady M, Sevastjanova R, Sperrle F, et al (2018) Progressive learning of topic modeling parameters: A visual analytics framework. IEEE Transactions on Visualization and Computer Graphics 24(1):382–391. https://doi.org/10.1109/tvcg.2017.2745080

65. El-Assady M, Sperrle F, Deussen O, et al (2019) Visual analytics for topic model optimization based on user-steerable speculative execution. IEEE Transactions on Visualization and Computer Graphics 25(1):374–384. https://doi.org/10.1109/tvcg.2018.2864769

66. El-Assady M, Kehlbeck R, Collins C, et al (2020) Semantic concept spaces: Guided topic model refinement using word-embedding projections. IEEE Transactions on Visualization and Computer Graphics 26(1):1001–1011. https://doi.org/10.1109/tvcg.2019.2934654

67. Elmqvist N, Dragicevic P, Fekete JD (2008) Rolling the dice: Multidimensional visual explo-ration using scatterplot matrix navigation. IEEE Transactions on Visualization and Computer Graphics 14(6):1539–1148

68. Eloundou T, Manning S, Mishkin P, et al (2023) Gpts are gpts: An early look at the labor market impact potential of large language models. arXiv preprint arXiv:2303.10130

69. van den Elzen S, van Wijk JJ (2011) BaobabView: Interactive construction and analysis of deci-sion trees. In: Proceedings of the IEEE Conference on Visual Analytics Science and Technology, pp 151–160. https://doi.org/10.1109/vast.2011.6102453

70. Engel PM, Heinen MR (2010) Incremental learning of multivariate gaussian mixture models. In: Proceedings of the Advances in Artificial Intelligence, pp 82–91

71. Fernstad SJ (2019) To identify what is not there: A definition of missingness patterns and evaluation of missing value visualization. Information Visualization 18(2):230–250

72. Fernstad SJ, Westberg JJ (2022) To explore what isn't there-glyph-based visualization for analy-sis of missing values. IEEE Transactions on Visualization and Computer Graphics 28(10):3513–3529

73. Ferreira N, Lins L, Fink D, et al (2011) BirdVis: Visualizing and understanding bird populations. IEEE Transactions on Visualization and Computer Graphics 17(12):2374–238. https://doi.org/10.1109/tvcg.2011.176

74. Fischer M, Kobs K, Hotho A (2020) NICER: Aesthetic image enhancement with humans in the loop. arXiv preprint arXiv:2012.01778

75. Foulds J, Frank E (2010) A review of multi-instance learning assumptions. The Knowledge Engineering Review 25(1):1–25

76. Fröhler B, Möller T, Heinzl C (2016) GEMSe: Visualization-guided exploration of multi-channel segmentation algorithms. Computer Graphics Forum 35(3):191–20. https://doi.org/10.1111/cgf.12895

77. Garcia Caballero HS, Westenberg MA, Gebre B, et al (2019) V-awake: A visual analytics approach for correcting sleep predictions from deep learning models. In: Computer Graphics Forum, pp 1–12

78. Gleicher M, Barve A, Yu X, et al (2020) Boxer: Interactive Comparison of Classifier Results. Computer Graphics Forum 39(3):181-193

79. Gomez O, Holter S, Yuan J, et al (2020) Vice: Visual counterfactual explanations for machine learning models. In: Proceedings of the International Conference on Intelligent User Interfaces, pp 531–535. https://doi.org/10.1145/3377325.3377536

80. Goodfellow I, Bengio Y, Courville A (2016) Deep learning. MIT press

81. Görtler J, Hohman F, Moritz D, et al (2022) Neo: Generalizing confusion matrix visualization to hierarchical and multi-output labels. In: Proceedings of the CHI Conference on Human Factors in Computing Systems. https://doi.org/10.1145/3491102.3501823

82. Gou L, Zou L, Li N, et al (2021) VATLD: A visual analytics system to assess, understand and improve traffic light detection. IEEE Transactions on Visualization and Computer Graphics 27(2):261–271

83. Gschwandtner T, Erhart O (2018) Know your enemy: Identifying quality problems of time series data. In: Proceedings of the IEEE Pacific Visualization Symposium, pp 205–214

84. Gschwandtner T, Aigner W, Miksch S, et al (2014) Timecleanser: A visual analytics approach for data cleansing of time-oriented data. In: Proceedings of the 14th International Conference on Knowledge Technologies and Data-Driven Business. Association for Computing Machinery. https://doi.org/10.1145/2637748.2638423

85. Guan C, Wang X, Zhang Q, et al (2019) Towards a deep and unified understanding of deep neural models in NLP. In: Proceedings of the International Conference on Machine Learning, pp 2454–2463

86. Gunning D, Aha D (2019) Darpa's explainable artificial intelligence (xai) program. AI magazine 40(2):44–58

87. Halter G, Ballester-Ripoll R, Flueckiger B, et al (2019) VIAN: A visual annotation tool for film analysis. Computer Graphics Forum 38(3):119–129

88. Han J, Wang C (2020) Ssr-tvd: Spatial super-resolution for time-varying data analysis and visualization. IEEE Transactions on Visualization and Computer Graphics 28(6):2445–2456

89. He J, Wang X, Wong KK, et al (2024) Videopro: A visual analytics approach for interactive video programming. IEEE Transactions on Visualization and Computer Graphics 30(1):87–9. https://doi.org/10.1109/TVCG.2023.3326586

90. He K, Sun J, Tang X (2010) Single image haze removal using dark channel prior. IEEE Transactions on Pattern Analysis and Machine Intelligence 33(12):2341–2353

91. He W, Lee TY, van Baar J, et al (2020) DynamicsExplorer: Visual analytics for robot control tasks involving dynamics and LSTM-based control policies. In: Proceedings of the IEEE Pacific Visualization Symposium, pp 36–45. https://doi.org/10.1109/pacificvis48177.2020.7127

92. He W, Zou L, Shekar AK, et al (2022) Where can we help? a visual analytics approach to diagnosing and improving semantic segmentation of movable objects. IEEE Transactions on Visualization and Computer Graphics 28(1):1040–1050

93. Heimerl F, Koch S, Bosch H, et al (2012) Visual classifier training for text document retrieval. IEEE Transactions on Visualization and Computer Graphics 18(12):2839–2848

94. Higgins I, Matthey L, Pal A, et al (2017) beta-VAE: Learning basic visual concepts with a constrained variational framework. In: International conference on learning representations

95. Hinterreiter A, Ruch P, Stitz H, et al (2022) Confusionflow: A model-agnostic visualization for temporal analysis of classifier confusion. IEEE Transactions on Visualization and Computer Graphics 28(2):1222–1236

96. Höferlin B, Netzel R, Höferlin M, et al (2012) Inter-active learning of ad-hoc classifiers for video visual analytics. In: Proceedings of the Conference on Visual Analytics Science and Technology, pp 23–32

97. Hoffman RR, Mueller ST, Klein G, et al (2018) Metrics for explainable ai: Challenges and prospects. arXiv preprint arXiv:1812.04608

98. Hohman F, Kahng M, Pienta R, et al (2019) Visual analytics in deep learning: An interrogative survey for the next frontiers. IEEE Transactions on Visualization and Computer Graphics 25(8):2674–2693

99. Hohman F, Park H, Robinson C, et al (2020) Summit: Scaling deep learning interpretability by visualizing activation and attribution summarizations. IEEE Transactions on Visualization and Computer Graphics 26(1):1096–1106. https://doi.org/10.1109/tvcg.2019.2934659

100. Hong F, Liu C, Yuan X (2019) DNN-VolVis: Interactive volume visualization supported by deep neural network. In: 2019 IEEE Pacific Visualization Symposium (PacificVis), IEEE, pp 282–291

101. Hong J, Zhu Z, Yu S, et al (2021) Federated adversarial debiasing for fair and transferable representations. In: Proceedings of the 27th ACM SIGKDD Conference on Knowledge Discovery & Data Mining, pp 617–627

102. Hong SR, Hullman J, Bertini E (2020) Human factors in model interpretability: Industry practices, challenges, and needs. Proceedings of the ACM on Human-Computer Interaction 4(CSCW1):1–2. https://doi.org/10.1145/3392878

103. Hoque MN, He W, Shekar AK, et al (2023) Visual concept programming: A visual analytics approach to injecting human intelligence at scale. IEEE Transactions on Visualization and Computer Graphics 29(1):74–83

104. Huang Y, Liu Y, Li C, et al (2019) Gbrtvis: online analysis of gradient boosting regression tree. Journal of Visualization 22:125–140

105. HuggingFace (2024) Huggingface. https://huggingface.co/models, Last accessed 2024-01-01
106. Ingram S, Munzner T, Irvine V, et al (2010) Dimstiller: Workflows for dimensional analysis and reduction. In: Proceedings of the IEEE Conference on Visual Analytics Science and Technology, pp 3–10
107. Jaunet T, Vuillemot R, Wolf C (2020) DRLViz: Understanding decisions and memory in deep reinforcement learning. Computer Graphics Forum 27(6):49-61
108. Jean CS, Ware C, Gamble R (2016) Dynamic change arcs to explore model forecasts. Computer Graphics Forum 35(3):311–32. https://doi.org/10.1111/cgf.12907
109. Jia S, Lin P, Li Z, et al (2020) Visualizing surrogate decision trees of convolutional neural networks. Journal of Visualization 23:141–156
110. Jia S, Li Z, Chen N, et al (2022) Towards visual explainable active learning for zero-shot classification. IEEE Transactions on Visualization and Computer Graphics 28(1):791–801
111. Jin Z, Wang Y, Wang Q, et al (2023) GNNLens: A visual analytics approach for prediction error diagnosis of graph neural networks. IEEE Transactions on Visualization and Computer Graphics 29(6):3024–3038
112. Johansson S, Johansson J (2009) Interactive dimensionality reduction through user-defined combinations of quality metrics. IEEE Transactions on Visualization and Computer Graphics 15(6):993–1000
113. Júnior AS, Renso C, Matwin S (2017) ANALYTiC: An active learning system for trajectory classification. IEEE Computer Graphics and Applications 37(5):28–39
114. Kahng M, Andrews PY, Kalro A, et al (2018) ActiVis: Visual exploration of industry-scale deep neural network models. IEEE Transactions on Visualization and Computer Graphics 24(1):88–9. https://doi.org/10.1109/tvcg.2017.2744718
115. Kahng M, Thorat N, Chau DHP, et al (2019) GAN lab: Understanding complex deep generative models using interactive visual experimentation. IEEE Transactions on Visualization and Computer Graphics 25(1):310–320. https://doi.org/10.1109/tvcg.2018.2864500
116. Kandel S, Parikh R, Paepcke A, et al (2012) Profiler: Integrated statistical analysis and visualization for data quality assessment. In: Proceedings of the International Working Conference on Advanced Visual Interfaces, pp 547–554
117. Karpathy A, Johnson J, Li FF (2015) Visualizing and understanding recurrent networks. CoRR abs/1506.02078. URL http://dblp.uni-trier.de/db/journals/corr/corr1506.html#KarpathyJL15
118. Kaul S, Borland D, Cao N, et al (2022) Improving visualization interpretation using counterfactuals. IEEE Transactions on Visualization and Computer Graphics 28(1):998–1008
119. Keim D, Andrienko G, Fekete JD, et al (2008) Visual analytics: Definition, process, and challenges. Springer
120. Khayat M, Karimzadeh M, Zhao J, et al (2020) VASSL: A visual analytics toolkit for social spambot labeling. IEEE Transactions on Visualization and Computer Graphics 26(1):874–883
121. Kim AY, Hardin J (2021) "playing the whole game": A data collection and analysis exercise with google calendar. Journal of Statistics and Data Science Education 29(sup1):S51–S60
122. Kim H, Drake B, Endert A, et al (2021) ArchiText: Interactive hierarchical topic modeling. IEEE Transactions on Visualization and Computer Graphics 27(9):3644-3655
123. Köksalan MM, Wallenius J, Zionts S (2011) Multiple criteria decision making: from early history to the 21st century. World Scientific
124. Krause J, Perer A, Bertini E (2014) INFUSE: Interactive feature selection for predictive modeling of high dimensional data. IEEE Transactions on Visualization and Computer Graphics 20(12):1614–1623
125. Krause J, Dasgupta A, Swartz J, et al (2017) A workflow for visual diagnostics of binary classifiers using instance-level explanations. In: Proceedings of the IEEE Conference on Visual Analytics Science and Technology, pp 162–172. https://doi.org/10.1109/vast.2017.8585720

126. Krizhevsky A, Hinton G (2009) Learning multiple layers of features from tiny images. Tech. Rep. 0, University of Toronto, Toronto, Ontario

127. Kurzhals K, Hlawatsch M, Seeger C, et al (2017) Visual analytics for mobile eye tracking. IEEE Transactions on Visualization and Computer Graphics 23(1):301–310

128. Kwon BC, Choi MJ, Kim JT, et al (2019) RetainVis: Visual analytics with interpretable and interactive recurrent neural networks on electronic medical records. IEEE Transactions on Visualization and Computer Graphics 25(1):299–30. https://doi.org/10.1109/tvcg.2018.2865027

129. Kwon BC, Anand V, Severson KA, et al (2021) DPVis: Visual analytics with hidden markov models for disease progression pathways. IEEE Transactions on Visualization and Computer Graphics 27(9):3685–3700

130. Kwon BC, Lee J, Chung C, et al (2022) DASH: Visual analytics for debiasing image classification via user-driven synthetic data augmentation. In: EuroVis 2022 - Short Paper. https://doi.org/10.2312/evs.20221099

131. Lakshminarayanan B, Pritzel A, Blundell C (2017) Simple and scalable predictive uncertainty estimation using deep ensembles. In: Proceedings of the Advances in Neural Information Processing Systems, pp 6402–6413

132. Langer M, Oster D, Speith T, et al (2021) What do we want from explainable artificial intelligence (XAI)?–a stakeholder perspective on XAI and a conceptual model guiding interdisciplinary XAI research. Artificial Intelligence 296:103473

133. LeCun Y, Bengio Y, Hinton G (2015) Deep learning. nature 521(7553):436–444

134. Lee H, Kihm J, Choo J, et al (2012) iVisClustering: An interactive visual document clustering via topic modeling. Computer Graphics Forum 31(3):1155–116. https://doi.org/10.1111/j.1467-8659.2012.03108.x

135. Lee S, Wang X, Han S, et al (2022) Self-explaining deep models with logic rule reasoning. arXiv preprint arXiv:2210.07024

136. Lefkimmiatis S (2018) Universal denoising networks: a novel cnn architecture for image denoising. In: Proceedings of the IEEE conference on computer vision and pattern recognition, pp 3204–3213

137. Lekschas F, Peterson B, Haehn D, et al (2020) PEAX: Interactive visual pattern search in sequential data using unsupervised deep representation learning. Computer Graphics Forum 39(3):167–179

138. Li G, Wang J, Shen HW, et al (2021a) CNNPruner: Pruning convolutional neural networks with visual analytics. IEEE Transactions on Visualization and Computer Graphics 27(2):1364–1373

139. Li Q, Njotoprawiro KS, Haleem H, et al (2018) EmbeddingVis: A visual analytics approach to comparative network embedding inspection. In: 2018 IEEE Conference on Visual Analytics Science and Technology (VAST), IEEE, pp 48–59

140. Li S, Araujo IB, Ren W, et al (2019) Single image deraining: A comprehensive benchmark analysis. In: Proceedings of the IEEE/CVF Conference on Computer Vision and Pattern Recognition, pp 3838–3847

141. Li Y, Wang J, Dai X, et al (2023a) How does attention work in vision transformers? a visual analytics attempt. IEEE Transactions on Visualization and Computer Graphics 29(6):2888–2900

142. Li Y, Wang J, Fujiwara T, et al (2023b) Visual analytics of neuron vulnerability to adversarial attacks on convolutional neural networks. ACM Transactions on Interactive Intelligent Systems 13(4):1-26. https://doi.org/10.1145/3587470

143. Li YF, Wang SB, Zhou ZH (2016) Graph quality judgement: A large margin expedition. In: Proceedings of the International Joint Conference on Artificial Intelligence, pp 1725–1731

144. Li YF, Guo LZ, Zhou ZH (2021b) Towards safe weakly supervised learning. IEEE Transactions on Pattern Analysis and Machine Intelligence 43(1):334–346

145. Li Z, Wang X, Yang W, et al (2022) A unified understanding of deep nlp models for text classification. IEEE Transactions on Visualization and Computer Graphics 28(12):4980–499. https://doi.org/10.1109/TVCG.2022.3184186

146. Liao QV, Varshney KR (2021) Human-centered explainable AI (XAI): From algorithms to user experiences. arXiv preprint arXiv:2110.10790

147. Liu M, Liu S, Zhu X, et al (2016) An uncertainty-aware approach for exploratory microblog retrieval. IEEE Transactions on Visualization and Computer Graphics 22(1):250–25. https://doi.org/10.1109/tvcg.2015.2467554

148. Liu M, Jiang L, Liu J, et al (2017a) Improving learning-from-crowds through expert validation. In: Proceedings of the International Joint Conference on Artificial Intelligence, pp 2329–2336

149. Liu M, Shi J, Li Z, et al (2017b) Towards better analysis of deep convolutional neural networks. IEEE Transactions on Visualization and Computer Graphics 23(1):91–10. https://doi.org/10.1109/tvcg.2016.2598831

150. Liu M, Liu S, Su H, et al (2018a) Analyzing the noise robustness of deep neural networks. In: IEEE Conference on Visual Analytics Science and Technology (VAST). IEE. https://doi.org/10.1109/vast.2018.8802509

151. Liu M, Shi J, Cao K, et al (2018b) Analyzing the training processes of deep generative models. IEEE Transactions on Visualization and Computer Graphics 24(1):77–8. https://doi.org/10.1109/tvcg.2017.2744938

152. Liu S (2023) Enhancing training data quality with visual analytics. Computer 56(11):4–6

153. Liu S, Cui W, Wu Y, et al (2014) A survey on information visualization: recent advances and challenges. The Visual Computer 30(12):1373–1393

154. Liu S, Wang X, Liu M, et al (2017c) Towards better analysis of machine learning models: A visual analytics perspective. Visual Informatics 1(1):48–56

155. Liu S, Andrienko G, Wu Y, et al (2018c) Steering data quality with visual analytics: The complexity challenge. Visual Informatics 2(4):191–197. https://doi.org/10.1016/j.visinf.2018.12.001

156. Liu S, Xiao J, Liu J, et al (2018d) Visual diagnosis of tree boosting methods. IEEE Transactions on Visualization and Computer Graphics 24(1):163–17. https://doi.org/10.1109/tvcg.2017.2744378

157. Liu S, Chen C, Lu Y, et al (2019a) An interactive method to improve crowdsourced annotations. IEEE Transactions on Visualization and Computer Graphics 25(1):235–245

158. Liu S, Li Z, Li T, et al (2019b) NLIZE: A perturbation-driven visual interrogation tool for analyzing and interpreting natural language inference models. IEEE Transactions on Visualization and Computer Graphics 25(1):651–66. https://doi.org/10.1109/tvcg.2018.2865230

159. Liu Z, Wang Y, Bernard J, et al (2022) Visualizing graph neural networks with corgie: Corresponding a graph to its embedding. IEEE Transactions on Visualization and Computer Graphics 28(6):2500–2516

160. Lowe T, Forster EC, Albuquerque G, et al (2016) Visual analytics for development and evaluation of order selection criteria for autoregressive processes. IEEE Transactions on Visualization and Computer Graphics 22(1):151–15. https://doi.org/10.1109/tvcg.2015.2467612

161. Lu J, Batra D, Parikh D, et al (2019) ViLBERT: Pretraining task-agnostic visiolinguistic representations for vision-and-language tasks. In: Proceedings of the Advances in Neural Information Processing Systems, pp 13–23

162. Ma KL (2009) In situ visualization at extreme scale: Challenges and opportunities. IEEE Computer Graphics and Applications 29(6):14–19

163. Ma Y, Maciejewski R (2021) Visual analysis of class separations with locally linear segments. IEEE Transactions on Visualization and Computer Graphics 27(1):241–253

164. Ma Y, Xie T, Li J, et al (2020) Explaining vulnerabilities to adversarial machine learning through visual analytics. IEEE Transactions on Visualization and Computer Graphics 26(1):1075–108. https://doi.org/10.1109/tvcg.2019.2934631

165. Ma Y, Fan A, He J, et al (2021) A visual analytics framework for explaining and diagnosing transfer learning processes. IEEE Transactions on Visualization and Computer Graphics 27(2):1385–1395

166. MacInnes J, Santosa S, Wright W (2010) Visual classification: Expert knowledge guides machine learning. IEEE Computer Graphics and Applications 30(1):8–1. https://doi.org/10.1109/mcg.2010.18

167. Marcus MP, Santorini B, Marcinkiewicz MA (1993) Building a large annotated corpus of English: The Penn Treebank. Computational Linguistics 19(2):313–330

168. McCallum AK, Nigam K, Rennie J, et al (2000) Automating the construction of internet portals with machine learning. Information Retrieval 3(2):127–163

169. Meng L, Van Den Elzen S, Vilanova A (2022) Modelwise: Interactive model comparison for model diagnosis, improvement and selection. In: Computer Graphics Forum, Wiley Online Library, pp 97–108

170. Migut M, Worring M (2010) Visual exploration of classification models for risk assessment. In: Proceedings of the IEEE Conference on Visual Analytics Science and Technology, pp 11–1. https://doi.org/10.1109/vast.2010.5652398

171. Migut M, van Gemert J, Worring M (2011) Interactive decision making using dissimilarity to visually represented prototypes. In: Proceedings of the IEEE Conference on Visual Analytics Science and Technology, pp 141–149. https://doi.org/10.1109/vast.2011.6102451

172. Ming Y, Cao S, Zhang R, et al (2017) Understanding hidden memories of recurrent neural networks. In: Proceedings of the IEEE Conference on Visual Analytics Science and Technology, pp 13–2. https://doi.org/10.1109/vast.2017.8585721

173. Ming Y, Qu H, Bertini E (2019a) RuleMatrix: Visualizing and understanding classifiers with rules. IEEE Transactions on Visualization and Computer Graphics 25(1):342–35. https://doi.org/10.1109/tvcg.2018.2864812

174. Ming Y, Xu P, Qu H, et al (2019b) Interpretable and steerable sequence learning via prototypes. In: Proceedings of the ACM SIGKDD International Conference on Knowledge Discovery & Data Mining, pp 903–91. https://doi.org/10.1145/3292500.3330908

175. Ming Y, Xu P, Cheng F, et al (2020) ProtoSteer: Steering deep sequence model with prototypes. IEEE Transactions on Visualization and Computer Graphics 26(1):238–24. https://doi.org/10.1109/tvcg.2019.2934267

176. Moehrmann J, Bernstein S, Schlegel T, et al (2011) Improving the usability of hierarchical representations for interactively labeling large image data sets. In: Proceedings of the International Conference on Human-Computer Interaction, pp 618–627

177. Morris C, Kriege NM, Bause F, et al (2020) Tudataset: A collection of benchmark datasets for learning with graphs. arXiv preprint arXiv:2007.08663

178. Mühlbacher T, Piringer H (2013) A partition-based framework for building and validating regression models. IEEE Transactions on Visualization and Computer Graphics 19(12):1962–1971

179. Mühlbacher T, Linhardt L, Möller T, et al (2018) TreePOD: Sensitivity-aware selection of pareto-optimal decision trees. IEEE Transactions on Visualization and Computer Graphics 24(1):174–183. https://doi.org/10.1109/tvcg.2017.2745158

180. Murugesan S, Malik S, Du F, et al (2019) Deepcompare: Visual and interactive comparison of deep learning model performance. IEEE Computer Graphics and Applications 39:47–59

181. Neto MP, Paulovich FV (2021) Explainable matrix - visualization for global and local interpretability of random forest classification ensembles. IEEE Transactions on Visualization and Computer Graphics 27(2):1427–1437. https://doi.org/10.1109/TVCG.2020.3030354

182. Ng A (2021) A chat with andrew on mlops: From model-centric to data-centric ai. https://www.youtube.com/watch?v=06-AZXmwHjo, Last accessed: 2021-05-21

183. Nie S, Healey C, Padia K, et al (2018) Visualizing deep neural networks for text analytics. In: Proceedings of the IEEE Pacific Visualization Symposium, pp 180–18. https://doi.org/10.1109/pacificvis.2018.00031

184. Northcutt CG, Athalye A, Mueller J (2021) Pervasive label errors in test sets destabilize machine learning benchmarks. arXiv preprint arXiv:2103.14749

185. Olson ML, Nguyen T, Dixit G, et al (2021) Contrastive identification of covariate shift in image data. In: Proceedings of the IEEE Visualization Conference, pp 36–4. https://doi.org/10.1109/VIS49827.2021.9623289

186. Ono JP, Castelo S, Lopez R, et al (2021) Pipelineprofiler: A visual analytics tool for the exploration of automl pipelines. IEEE Transactions on Visualization and Computer Graphics 27(2):390–400

187. OpenAI (2023) Gpt-4 technical report. 2303.08774

188. Ouyang L, Wu J, Jiang X, et al (2022) Training language models to follow instructions with human feedback. Advances in Neural Information Processing Systems 35:27730–27744

189. Packer E, Bak P, Nikkila M, et al (2013) Visual analytics for spatial clustering: Using a heuristic approach for guided exploration. IEEE Transactions on Visualization and Computer Graphics 19(12):2179–2188. https://doi.org/10.1109/tvcg.2013.224

190. Paiva JGS, Schwartz WR, Pedrini H, et al (2015) An approach to supporting incremental visual data classification. IEEE Transactions on Visualization and Computer Graphics 21(1):4–17

191. Palmeiro J, Malveiro B, Costa R, et al (2022) Data+Shift: Supporting Visual Investigation of Data Distribution Shifts by Data Scientists. In: Proceedings of the Eurographics Conference on Visualization. https://doi.org/10.2312/evs.20221097

192. Park C, Lee J, Han H, et al (2019) Comdia+: An interactive visual analytics system for comparing, diagnosing, and improving multiclass classifiers. In: 2019 IEEE Pacific Visualization Symposium (PacificVis), IEEE, pp 313–317

193. Park JH, Nadeem S, Mirhosseini S, et al (2016) C^2A: Crowd consensus analytics for virtual colonoscopy. In: Proceedings of the IEEE Conference on Visual Analytics Science and Technology, pp 21–30

194. Park JH, Nadeem S, Boorboor S, et al (2021) CMed: Crowd analytics for medical imaging data. IEEE Transactions on Visualization and Computer Graphics 27(6):2869-2880

195. Pezzotti N, Lelieveldt BP, Van Der Maaten L, et al (2017) Approximated and user steerable tsne for progressive visual analytics. IEEE Transactions on Visualization and Computer Graphics 23(7):1739–1752

196. Pezzotti N, Hollt T, Gemert JV, et al (2018) DeepEyes: Progressive visual analytics for designing deep neural networks. IEEE Transactions on Visualization and Computer Graphics 24(1):98–108. https://doi.org/10.1109/tvcg.2017.2744358

197. Piringer H, Berger W, Krasser J (2010) HyperMoVal: Interactive visual validation of regression models for real-time simulation. Computer Graphics Forum 29(3):983–99. https://doi.org/10.1111/j.1467-8659.2009.01684.x

198. Pleiss G, Raghavan M, Wu F, et al (2017) On fairness and calibration. Advances in neural information processing systems 30

199. Pühringer M, Hinterreiter A, Streit M (2020) Instanceflow: Visualizing the evolution of classifier confusion at the instance level. In: 2020 IEEE visualization conference (VIS), IEEE, pp 291–295

200. Radford A, Kim JW, Hallacy C, et al (2021) Learning transferable visual models from natural language supervision. In: International Conference on Machine Learning

201. Raffel C, Shazeer N, Roberts A, et al (2020) Exploring the limits of transfer learning with a unified text-to-text transformer. The Journal of Machine Learning Research 21(1):5485–5551

202. Rapp T, Peters C, Dachsbacher C (2022) Image-based visualization of large volumetric data using moments. IEEE Transactions on Visualization and Computer Graphics 28(6):2314–2325

203. Rauber PE, Fadel SG, Falcao AX, et al (2017) Visualizing the hidden activity of artificial neural networks. IEEE Transactions on Visualization and Computer Graphics 23(1):101–11. https://doi.org/10.1109/tvcg.2016.2598838

204. Ren D, Amershi S, Lee B, et al (2017) Squares: Supporting interactive performance analysis for multiclass classifiers. IEEE Transactions on Visualization and Computer Graphics 23(1):61–70. https://doi.org/10.1109/tvcg.2016.2598828

205. Ribeiro MT, Singh S, Guestrin C (2016) "why should i trust you?": Explaining the predictions of any classifier. In: Proceedings of the 22nd ACM SIGKDD International Conference on Knowledge Discovery and Data Mining, pp 1135—114. https://doi.org/10.1145/2939672.2939778

206. Richer G, Pister A, Abdelaal M, et al (2022) Scalability in visualization. IEEE Transactions on Visualization and Computer Graphics. https://doi.org/10.1109/TVCG.2022.3231230, to be published

207. Roels J, Vernaillen F, Kremer A, et al (2019) A "human-in-the-loop" approach for semi-automated image restoration in electron microscopy. bioRxiv https://doi.org/10.1101/644146, URL https://www.biorxiv.org/content/early/2019/06/04/644146, https://www.biorxiv.org/content/early/2019/06/04/644146.full.pdf

208. Roesch I, Günther T (2019) Visualization of neural network predictions for weather forecasting. In: Computer Graphics Forum, Wiley Online Library, pp 209–220

209. Rohlig M, Luboschik M, Kruger F, et al (2015) Supporting activity recognition by visual analytics. In: Proceedings of the IEEE Conference on Visual Analytics Science and Technology, pp 41–4. https://doi.org/10.1109/vast.2015.7347629

210. Rooij O, van Wijk J, Worring M (2010) MediaTable: Interactive categorization of multimedia collections. IEEE Computer Graphics and Applications 30(5):42–51

211. Ropinski T, Oeltze S, Preim B (2011) Survey of glyph-based visualization techniques for spatial multivariate medical data. Computers & Graphics 35(2):392–401

212. Russakovsky O, Deng J, Su H, et al (2015) ImageNet Large Scale Visual Recognition Challenge. International Journal of Computer Vision 115(3):211–252

213. Sacha D, Kraus M, Bernard J, et al (2018) SOMFlow: Guided exploratory cluster analysis with self-organizing maps and analytic provenance. IEEE Transactions on Visualization and Computer Graphics 24(1):120–130. https://doi.org/10.1109/tvcg.2017.2744805

214. Sánchez A, Soguero-Ruíz C, Mora-Jiménez I, et al (2018) Scaled radial axes for interactive visual feature selection: A case study for analyzing chronic conditions. Expert Systems with Applications 100:182–196

215. Sarah G, et al (2022) Introduction to machine learning with python. Oreilly

216. Scheepens R, Michels S, van de Wetering H, et al (2015) Rationale visualization for safety and security. Computer Graphics Forum 34(3):191–200.https://doi.org/10.1111/cgf.12631

217. Schultz T, Kindlmann GL (2013) Open-box spectral clustering: Applications to medical image analysis. IEEE Transactions on Visualization and Computer Graphics 19(12):2100–210. https://doi.org/10.1109/tvcg.2013.181

218. Schwartz R, Dodge J, Smith NA, et al (2020) Green AI. Communications of the ACM 63(12):54–63

219. Sharifi Noorian S, Qiu S, Sayin B, et al (2023) Perspective: Leveraging human understanding for identifying and characterizing image atypicality. In: Proceedings of the International Conference on Intelligent User Interfaces. Association for Computing Machinery, New York, NY, USA, IUI '23, p 650-663. https://doi.org/10.1145/3581641.3584096

220. Shen Q, Wu Y, Jiang Y, et al (2020) Visual interpretation of recurrent neural network on multi-dimensional time-series forecast. In: Proceedings of the IEEE Pacific Visualization Symposium, pp 61–70. https://doi.org/10.1109/pacificvis48177.2020.2785

221. Smilkov D, Carter S, Sculley D, et al (2017) Direct-manipulation visualization of deep networks. In: ICML Workshop on Vis for Deep Learning

222. Snyder LS, Lin YS, Karimzadeh M, et al (2020) Interactive learning for identifying relevant tweets to support real-time situational awareness. IEEE Transactions on Visualization and Computer Graphics 26(1):558–568

223. Sperrle F, Sevastjanova R, Kehlbeck R, et al (2019) VIANA: Visual interactive annotation of argumentation. In: Proceedings of the Conference on Visual Analytics Science and Technology, pp 11–22

224. Spinner T, Schlegel U, Schafer H, et al (2020) explAIner: A visual analytics framework for interactive and explainable machine learning. IEEE Transactions on Visualization and Computer Graphics 26(1):1064–1074. https://doi.org/10.1109/tvcg.2019.2934629

225. Stein M, Janetzko H, Breitkreutz T, et al (2016) Director's cut: Analysis and annotation of soccer matches. IEEE Computer Graphics and Applications 36(5):50–60

226. Stolper CD, Perer A, Gotz D (2014) Progressive visual analytics: User-driven visual exploration of in-progress analytics. IEEE Transactions on Visualization and Computer Graphics 20(12):1653–1662

227. Strobelt H, Gehrmann S, Pfister H, et al (2018) LSTMVis: A tool for visual analysis of hidden state dynamics in recurrent neural networks. IEEE Transactions on Visualization and Computer Graphics 24(1):667–676. https://doi.org/10.1109/tvcg.2017.2744158

228. Strobelt H, Gehrmann S, Behrisch M, et al (2019) Seq2seq-Vis: A visual debugging tool for sequence-to-sequence models. IEEE Transactions on Visualization and Computer Graphics 25(1):353–363. https://doi.org/10.1109/tvcg.2018.2865044

229. Strobelt H, Kinley J, Krueger R, et al (2022) Genni: Human-ai collaboration for data-backed text generation. IEEE Transactions on Visualization and Computer Graphics 28(1):1106–1116

230. Sun M, Mi P, North C, et al (2016) Biset: Semantic edge bundling with biclusters for sense-making. IEEE Transactions on Visualization and Computer Graphics 22(1):310–319

231. Tam GK, Fang H, Aubrey AJ, et al (2011) Visualization of time-series data in parameter space for understanding facial dynamics. Computer Graphics Forum 30(3):901–910

232. Teoh ST, Ma KL (2003) Paintingclass: interactive construction, visualization and exploration of decision trees. In: Proceedings of the ninth ACM SIGKDD international conference on Knowledge discovery and data mining, pp 667–672

233. Turkay C, Filzmoser P, Hauser H (2011) Brushing dimensions-a dual visual analysis model for high-dimensional data. IEEE Transactions on Visualization and Computer Graphics 17(12):2591–2599

234. Tzeng FY, Ma KL (2005) Opening the black box - data driven visualization of neural networks. In: Proceedings of the IEEE Conference on Visualization, pp 383–39. https://doi.org/10.1109/visual.2005.1532820

235. Ulyanov D, Vedaldi A, Lempitsky V (2018) Deep image prior. In: Proceedings of the IEEE conference on computer vision and pattern recognition, pp 9446–9454

236. Vaswani A, Shazeer N, Parmar N, et al (2017) Attention is all you need. Advances in neural information processing systems 30

237. Vrotsou K, Nordman A (2019) Exploratory visual sequence mining based on pattern-growth. IEEE Transactions on Visualization and Computer Graphics 25(8):2597–261. https://doi.org/10.1109/tvcg.2018.2848247

238. Wale N, Watson IA, Karypis G (2008) Comparison of descriptor spaces for chemical compound retrieval and classification. Knowledge and Information Systems 14(3):347–375

239. Wang J, Gou L, Yang H, et al (2018) Ganviz: A visual analytics approach to understand the adversarial game. IEEE Transactions on Visualization and Computer Graphics 24(6):1905–191. https://doi.org/10.1109/tvcg.2018.2816223

240. Wang J, Gou L, Shen HW, et al (2019a) DQNViz: A visual analytics approach to understand deep q-networks. IEEE Transactions on Visualization and Computer Graphics 25(1):288–29. https://doi.org/10.1109/tvcg.2018.2864504

241. Wang J, Gou L, Zhang W, et al (2019b) DeepVID: Deep visual interpretation and diagnosis for image classifiers via knowledge distillation. IEEE Transactions on Visualization and Computer Graphics 25(6):2168–218. https://doi.org/10.1109/tvcg.2019.2903943

242. Wang J, Zhang W, Yang H (2020a) SCANViz: Interpreting the symbol-concept association captured by deep neural networks through visual analytics. In: Proceedings of the IEEE Pacific Visualization Symposium, pp 51–6. https://doi.org/10.1109/pacificvis48177.2020.3542

243. Wang J, Zhang W, Wang L, et al (2021a) Investigating the evolution of tree boosting models with visual analytics. In: IEEE Pacific Visualization Symposium (PacificVis), IEEE, pp 186–195

244. Wang J, Zhang W, Yang H, et al (2022a) Visual analytics for rnn-based deep reinforcement learning. IEEE Trans Vis Comput Graph 28(12):4141–4155

245. Wang Q, Yuan J, Chen S, et al (2020b) Visual genealogy of deep neural networks. IEEE Transactions on Visualization and Computer Graphics 26(11):3340–3352

246. Wang Q, Xu Z, Chen Z, et al (2021b) Visual analysis of discrimination in machine learning. IEEE Transactions on Visualization and Computer Graphics 27(2):1470–1480

247. Wang W, Dai J, Chen Z, et al (2022b) InternImage: Exploring large-scale vision foundation models with deformable convolutions. arXiv preprint arXiv:2211.05778

248. Wang X, Liu S, Liu J, et al (2016) TopicPanorama: A full picture of relevant topics. IEEE Transactions on Visualization and Computer Graphics 22(12):2508–252. https://doi.org/10.1109/tvcg.2016.2515592

249. Wang X, Chen W, Xia J, et al (2020c) ConceptExplorer: Visual analysis of concept drifts in multi-source time-series data. In: IEEE Conference on Visual Analytics Science and Technology (VAST). Institute of Electrical and Electronics Engineers (IEEE), pp 13–24

250. Wang X, He J, Jin Z, et al (2022c) M2Lens: visualizing and explaining multimodal models for sentiment analysis. IEEE Transactions on Visualization and Computer Graphics 28(1):802–812

251. Wang X, Chen W, Xia J, et al (2023) Hetvis: A visual analysis approach for identifying data heterogeneity in horizontal federated learning. IEEE Transactions on Visualization and Computer Graphics 29(1):310–319

252. Wang Y, Li J, Nie F, et al (2017) Linear discriminative star coordinates for exploring class and cluster separation of high dimensional data. In: Computer Graphics Forum, Wiley Online Library, pp 401–410

253. Weber T, Hußmann H, Han Z, et al (2020) Draw with me: Human-in-the-loop for image restoration. In: Proceedings of the International Conference on Intelligent User Interfaces, New York, NY, USA, p 243-253. https://doi.org/10.1145/3377325.3377509

254. Wexler J, Pushkarna M, Bolukbasi T, et al (2020) The what-if tool: Interactive probing of machine learning models. IEEE Transactions on Visualization and Computer Graphics 26(1):56–6. https://doi.org/10.1109/tvcg.2019.2934619

255. Willett W, Heer J, Agrawala M (2007) Scented widgets: Improving navigation cues with embedded visualizations. IEEE Transactions on Visualization and Computer Graphics 13(6):1129–1136

256. Wongsuphasawat K, Smilkov D, Wexler J, et al (2018) Visualizing dataflow graphs of deep learning models in TensorFlow. IEEE Transactions on Visualization and Computer Graphics 24(1):1–1. https://doi.org/10.1109/tvcg.2017.2744878

257. Wu Y, Cui W, Song Y, et al (2012) A survey on topic-based text visual analytics. Journal of Computer-Aided Design & Computer Graphics 24(10):1266–1272

258. Xenopoulos P, Rulff J, Nonato LG, et al (2023) Calibrate: Interactive analysis of probabilistic model output. IEEE Transactions on Visualization and Computer Graphics 29(1):853–86. https://doi.org/10.1109/TVCG.2022.3209489

259. Xia J, Li J, Chen S, et al (2021) A survey on interdisciplinary research of visualization and artificial intelligence. SCIENTIA SINICA Informationis 51(11):1777–1791. https://doi.org/10. 1360/SSI-2021-0062, URL http://www.sciengine.com/publisher/ScienceChinaPress/journal/ SCIENTIASINICAInformationis/51/11/10.1360/SSI-2021-0062

260. Xiang S, Ye X, Xia J, et al (2019) Interactive correction of mislabeled training data. In: Proceedings of the IEEE Conference on Visual Analytics Science and Technology, pp 57–68

261. Xiao J, Liu M, Liu S (2016) A visual analysis system for news data. Journal of Computer-Aided Design & Computer Graphics 28(11):1863–1871. URL https://www.jcad.cn/cn/article/ id/72d2b625-4463-4f9f-a06c-6f23e00816b7

262. Xie T, Ma Y, Kang J, et al (2022) Fairrankvis: A visual analytics framework for exploring algorithmic fairness in graph mining models. IEEE Transactions on Visualization and Computer Graphics 28(1):368–377. https://doi.org/10.1109/TVCG.2021.3114850

263. Yang W, Li Z, Liu M, et al (2020) Diagnosing concept drift with visual analytics. In: IEEE Conference on Visual Analytics Science and Technology (VAST). Institute of Electrical and Electronics Engineers (IEEE), pp 1–1. https://doi.org/10.1109/vast50239.2020.00007

264. Yang W, Wang X, Lu J, et al (2021) Interactive steering of hierarchical clustering. IEEE Transactions on Visualization and Computer Graphics 27(10):3953–3967

265. Yang W, Ye X, Zhang X, et al (2022) Diagnosing ensemble few-shot classifiers. IEEE Transactions on Visualization and Computer Graphics 28(9):3292–3306. https://doi.org/10.1109/ TVCG.2022.3182488

266. Yang W, Chen C, Zhu J, et al (2023) A survey of visual analytics research for improving training data quality. Journal of Computer-Aided Design & Computer Graphics 35(11):1629–164. https://doi.org/10.3724/SP.J.1089.2023.2023-00321

267. Yang W, Guo Y, Wu J, et al (2024a) Interactive reweighting for mitigating label quality issues. IEEE Transactions on Visualization and Computer Graphics 30(3):1837–1852

268. Yang W, Liu M, Wang Z, et al (2024b) Foundation models meet visualizations: Challenges and opportunities. Computational Visual Media. arXiv:2310.05771

269. Ying Z, Bourgeois D, You J, et al (2019) Gnnexplainer: Generating explanations for graph neural networks. Advances in neural information processing systems 32

270. Yuan J, Chen C, Yang W, et al (2021) A survey of visual analytics techniques for machine learning. Computational Visual Media 7(1):3–36

271. Yuan J, Liu M, Tian F, et al (2023) Visual analysis of neural architecture spaces for summarizing design principles. IEEE Transactions on Visualization and Computer Graphics 29(1):288–298

272. Zeiler MD, Fergus R (2014) Visualizing and understanding convolutional networks. In: Proceedings of the European Conference on Computer Vision, pp 818–833

273. Zgraggen E, Galakatos A, Crotty A, et al (2017) How progressive visualizations affect exploratory analysis. IEEE Transactions on Visualization and Computer Graphics 23(8):1977–1987

274. Zhang C, Yang J, Zhan FB, et al (2016) A visual analytics approach to high-dimensional logistic regression modeling and its application to an environmental health study. In: Proceedings of the IEEE Pacific Visualization Symposium, pp 136–14. https://doi.org/10.1109/pacificvis.2016. 7465261

275. Zhang J, Wang Y, Molino P, et al (2019) Manifold: A model-agnostic framework for interpretation and diagnosis of machine learning models. IEEE Transactions on Visualization and Computer Graphics 25(1):364–373. https://doi.org/10.1109/tvcg.2018.2864499

276. Zhang X, Liu Z, Yang W, et al (2022) The more, the better? active silencing of non-positive transfer for efficient multi-domain few-shot classification. In: Proceedings of the 30th ACM International Conference on Multimedia. Association for Computing Machinery, New York, NY, USA, MM '22, pp 1993–200. https://doi.org/10.1145/3503161.3548349

277. Zhao J, Karimzadeh M, Masjedi A, et al (2019a) FeatureExplorer: Interactive feature selection and exploration of regression models for hyperspectral images. In: 2019 IEEE Visualization Conference (VIS), IEEE, pp 161–165

278. Zhao K, Ward MO, Rundensteiner EA, et al (2014) LoVis: Local pattern visualization for model refinement. Computer Graphics Forum 33(3):331–340. https://doi.org/10.1111/cgf.12389

279. Zhao X, Wu Y, Lee DL, et al (2019b) iForest: Interpreting random forests via visual analytics. IEEE Transactions on Visualization and Computer Graphics 25(1):407–41. https://doi.org/10.1109/tvcg.2018.2864475

280. Zhou C, Liu P, Xu P, et al (2023) LIMA: Less is more for alignment. arXiv preprint arXiv:2305.11206

281. Zhou ZH (2006) Multi-instance learning from supervised view. Journal of Computer Science and Technology 21(5):800–809

282. Zhou ZH (2018) A brief introduction to weakly supervised learning. National Science Review 5(1):44–53

283. Zhou ZH, Tan ZH (2022) Learnware: Small models do big. arXiv preprint arXiv:2210.03647

284. Zhu M, Pan P, Chen W, et al (2020) EEMEFN: Low-light image enhancement via edge-enhanced multi-exposure fusion network. In: Proceedings of the AAAI Conference on Artificial Intelligence, pp 13106–13113